The Political Economy of Mobile Telephony in South Africa

The Political Economy of Mobile Telephony in South Africa

MTN, Vodacom and the State

Odilile Ayodele

ZED

LONDON • NEW YORK • OXFORD • NEW DELHI • SYDNEY

Zed
Bloomsbury Publishing Plc
50 Bedford Square, London, WC1B 3DP, UK
1385 Broadway, New York, NY 10018, USA
29 Earlsfort Terrace, Dublin 2, Ireland

BLOOMSBURY, Zed and the Zed logo are trademarks of Bloomsbury Publishing Plc

First published in Great Britain 2024

Series design by Adriana Brioso
Cover image © Grobler du Preez/Alamy Stock Photo

A catalogue record for this book is available from the British Library.

A catalog record for this book is available from the Library of Congress.

ISBN: HB: 978-1-3504-1177-7
ePDF: 978-1-3504-1179-1
eBook: 978-1-3504-1178-4

Typeset by Newgen KnowledgeWorks Pvt. Ltd., Chennai, India

BLOOMSBURY
OPEN ACCESS

An ebook edition of this book is available open access on bloomsburycollections.com. Open access was funded by the Bloomsbury Open Collections Library Collective.

Bloomsbury Open Collections is a collective-action approach to funding open access books that allows select authors to publish their books open access at no cost to them. Through this model, we make open access publication available to a wider range of authors by spreading the cost across multiple organizations, while providing additional benefits to participating libraries. The aim is to engage a more diverse set of authors and bring their work to a wider global audience.

More details, including how to participate and a list of contributing libraries, are available from http://www.bloomsbury.com/bloomsbury-open-collections.

Contents

Illustrations

Figures

Tables

Preface

The idea for this book began with my graduate studies when I realized the level of power mobile telephony companies wield in the Global South. At the time, many scholars and commentators focused on how these companies were a proxy for the newly democratic South Africa in their quest to build new relationships across the continent. Yet my research found the relationship between mobile telephony companies and the state – particularly in South Africa – is an intricate web of relations and interests.

The first iteration of this book began as my doctoral research. I asked whether or not mobile telephony network operators (MNOs) were an instrument of South Africa's foreign economic policy on the African continent. This book builds on this research and considers the power of mobile telephony companies in two contexts: as a transformational agent and how companies become embedded in foreign policy.

MNOs have shifted from mere telephony services providers as technology has become more sophisticated; they are also network providers, platform operators, data companies and financial service providers. Their evolution intensifies their importance as a potential cure-all for Africa's problems.

History shows that communications technology is an important diplomatic, economic and political tool. Technology companies can act as transformational agents in different spheres of society. The transformational power of mobile telephony operations gives mobile telephony corporations privileged access to elites in their host nation and home country. The novelty of mobile telephony, and the desperate need to increase access to communications, has meant that it cannot be viewed outside the prism of politics in African countries. The ability to ensure that the population has access to modes of communication is not only a commercial concern, but it is also a development concern, therefore putting it squarely under the ambit of politics. Every aspect of mobile telephony is affected by political interest, from the regulatory aspects of licensing of mobile operators to the shareholding of mobile telephony operators. The politicization of the sector is evident everywhere, including South Africa, whose foray into mobile telephony closely paralleled the political transformations within the country. The transition towards an inclusive democracy, the evolution of South Africa's corporate sector and the change in the country's relationships outside its borders all makeup part of the intricate story of South Africa's political economy of mobile telephony.

The multi-layered dynamics of mobile telephony transcend national boundaries. They speak to the ever-evolving relations between business and state, the importance of communications technology to development and the implications of increased mobile teledensity. The increasing evidence of mobile telephony's impact on economic growth and poverty alleviation is a testament to its transformative potential (Aker and Mbiti, 2010; Odhiambo, 2022).

Large telecommunications corporations worldwide, including Vodafone and Deutsche Telekom, have grappled with the complexities of shifting political agendas in their home countries and the evolving regulatory landscape surrounding mobile telephony. It is not an isolated trend but a reflection of the broader global developments in state-business relations and the complex interplay of political, economic and social forces that shape the development and deployment of mobile telephony.

South Africa's narrative is particularly compelling, weaving together three intriguing strands: the interplay between politics and business, the intersection of politics and technology in a developing country context and the convergence of politics and communication technologies. These threads mirror both global trends, regional priorities and local evolutions, highlighting the intricate dance between external influences and local contexts in shaping the trajectory of mobile telephony.

In this book, I look at the close relationship between commerce, technology and the extension of the state's power inside and outside its borders. I unpack this transformational power under the rubric of political economy and trace the growing influence of South African-originated mobile telephony companies. Furthermore, I highlight the thin line between corporatism and harnessing emerging technologies as an embodiment of interstate political power.

The genesis of MNOs in South Africa's case runs parallel to the country's democratic transition, providing an opportunity to garner several lessons on how contemporary technology companies become embedded in foreign policy orientation – from an African perspective. In the book, I underscore South Africa's post-apartheid corporate expansion into the African continent, underpinned by neoliberal ideologies. Moreover, it offers a distinctive analytical outlook into the role of multinational corporations (MNCs) in this dynamic, particularly those that are partly homegrown, different from the usual focus on industries such as mining, energy and manufacturing.

I trace the history of mobile telephony in South Africa, focusing on the primary operators, Vodacom and MTN. Tracing this history, one can observe a complex interplay of state intervention, foreign investment, market regulation and local political realities. For instance, in the Vodacom case, we see the evolving state-society-business relationship in the context of political transformation. Both the MTN and Vodacom narratives allow us to observe continuity and change in the state-business-society relationship. We can then explain how these dynamics shape the development and governance of the sector in South Africa and the rest of the country.

Odilile Ayodele
September 2024

Acronyms

ANC	African National Congress
APTU	African Postal and Telecommunications Union
AsgiSA	Accelerated Shared Growth and Investment
ATTC	African Transcontinental Telegraph Company
ATU	African Telecommunications Union
AU	African Union
BBBEE	Broad-Based Black Economic Empowerment
BEE	Black economic empowerment
BEEG	Black Economic Empowerment Group
BRICS	Brazil, Russia, India, China and South Africa
BSAC	British South Africa Company
BT	British Telecom
BUSA	Business Unity South Africa
CAR	Central African Republic
CODESA	Convention for Democratic South Africa
COM	Chamber of Mines
COMESA	Common Market for Eastern and Southern Africa
CONESTEL	Collectif de opérateur nationaux exploitant dans le secteur des télécommunications
CONSAS	Constellation of Southern African States
COSATU	Congress of South African Trade Unions
CSI	corporate social investment
CSIR	Council for Scientific and Industrial Research (South Africa)
CSO	civil society organization
CSR	corporate social responsibility
DBSA	Development Bank of South Africa
DDG	deputy director-general
DEIC	Dutch East India Company

DFA	Department of Foreign Affairs
DG	director-general
DIRCO	Department of International Relations and Cooperation
DRC	Democratic Republic of the Congo
DTPS	Department of Posts and Telecommunications
EAC	East Africa Community
ECA	Electronic Communications Act
ECOSOCC	Economic, Social and Cultural Council
ECSC	European Coal and Steel Community
ECT	Electronic Communications and Transactions Act
EEIC	English East India Company
EPA	Economic Partnership Agreement
ERT	European Roundtable of Industrialists
ESCOM	Electricity Supply Commission
ESKOM	merger between ESCOM and EVKOM
EU	European Union
EVKOM	Elektrisiteitsvoorsieningskommissie
FCI	Federated Chamber of Industry
FDI	Foreign Direct Investment
FEDUSA	Federation of Unions in South Africa
FOCAC	Forum for China-Africa Cooperation
GDP	gross domestic product
GEAR	Growth, Employment and Redistribution Strategy
GPRS	General Packet Radio Service
GSM	Global System for Mobile Communications
GSMA	Global System for Mobile Communications Association
GTB	Guaranty Trust Bank
HCI	Hosken Consolidated Investments
ICASA	Independent Communications Authority of South Africa
ICT	Information and Communications Technology
IDC	Industrial Development Corporation
IFC	International Finance Corporation
IPOB	Indigenous People of Biafra
IR	International Relations
ISCOR	Iron and Steel Corporation

IT	information technology
ITU	International Telecommunications Union
JSE	Johannesburg Stock Exchange
MDGs	Millennium Development Goals
MENA	Middle East and North Africa
MIC	Mineworkers Investment Company
MNC	multinational corporation
MNO	mobile network operator
MTBPS	Medium Term Budget Policy Statement
MTN	Mobile Telephone Network
NACTU	National Council of Trade Unions
NAI	New Africa Investments
NAIL	New Africa Investment Limited
NAM	Non-Alignment Movement
NCC	Nigerian Communications Commission
NDP	National Development Plan
NEC	National Economic Forum
NEDLAC	National Economic Development and Labour Council
NEF	National Empowerment Fund
NEPAD	New Partnership for Africa's Development
NGO	non-governmental organization
NGP	New Growth Path
NMC	National Manpower Commission
NP	National Party
OAU	Organisation of African Unity
OECD	Organisation for Economic Cooperation and Development
Panaftel	Pan-African Telecommunications Network
PIC	public investment corporation
PICI	Presidential Infrastructure Champion Initiative
POTWA	Postal and Telecommunications Workers Association
PPP	public-private partnership
RASCOM	Regional African Satellite Communications Organisation
RDP	Reconstruction and Development Programme
RPF	Rwanda Patriotic Front
SACP	South African Communist Party

SACTWU	South African Clothing and Textile Workers Union
SACU	Southern African Customs Union
SADC	Southern African Development Community
SADCC	Southern African Development Coordination Conference
SANDF	South African National Defence Force
SAPT	South African Posts and Telecommunications
SARS	South African Revenue Service
SATRA	South African Telecommunications Regulatory Authority
SBC	small business corporation
SBR	state business relations
SCFA	Sub-Council on Foreign Affairs
SENS	Stock Exchange News Service
SME	small and medium-sized enterprise
SMS	short message service
SOE	state-owned enterprise
TCC	transnational capitalist class
TRIPS	Trade-Related Aspects of Intellectual Property Rights
UN	United Nations
UNCAC	United Nations Convention against Corruption
UNDP	United Nations Development Programme
UNESCO	United Nations Educational, Scientific and Cultural Organization
UNIDO	United Nations Industrial Development Organization
UNSC	United Nations Security Council
UNSG	United Nations Secretary-General
US	United States
USA	Universal Service Agency
USO	Universal Service Obligations
WIPHOLD	Women Investment Portfolio Holdings
WTO	World Trade Organization

Introduction

The year 1993 can easily be described as the twilight of South Africa's political and economic rebirth. Despite the bloodshed that marked that year, significant progress was made towards South Africa's democracy and its political economy. Significant major developments were the restoration of South African citizenship to those excluded due to the apartheid's homeland system, and the endorsement of the interim constitution, which paved the way for a fully inclusive election the following year. Against this backdrop, South Africa's telecommunications landscape was also undergoing intensive transformation. As the United Nations General Assembly ended the sanctions against South Africa, the Global System for Mobile Communications (GSM) debuted. This book tells of the political economy of mobile telephony in South Africa – it is a story that goes beyond economic concerns but is mired in politics and the metamorphosis of society. From the first GSM signal to the creation of pan-African mobile telecommunication giants, Mobile Telephone Network (MTN) and Vodacom have played an instrumental role in South Africa's domestic politics as well as the growth of mobile telephony on the African continent. The economic and social behaviour of the new companies were embedded in the bourgeoning state formation – by design. The birth of MTN and Vodacom runs parallel to the one of the nascent South African democratic government.

Apartheid meant many Black people were not included in plans for access to communication and hence got left behind. The ability to communicate is thus seen as a development pathway in the new South Africa. There was an awareness in the incoming administration that there were not enough resources to ensure that communication technology was distributed widely. Moreover, the new government also understood that the state needed to transform the face of South Africa's business to one that was socially responsive and that white men did not dominate.

The Political Economy of Mobile Telephony in South Africa: MTN, Vodacom and the State explores the intersection of South Africa's corporate expansion, mobile telephony and the economic dimension of South Africa's foreign policy. It focuses on the evolution of the relationship between corporate interests and South Africa's foreign policy, with the political transformations in the country as its cornerstone. It also provides an additional dimension to exploring South Africa's economic role on the African continent. Three distinct lines of enquiry undergird this book:

1. What political and economic gap does mobile telephony fill in South Africa?
2. What significant shifts have influenced private and public relationships between business, society and the state?
3. How has the change in South Africa's position regionally and internationally affected the growth of the private telephony industry and its primary business stakeholders?

This book draws together the fields of international relations, African studies, political economy, communication studies, history of technology, history of South African business and business studies. It goes beyond the mere focus on organizations or profiling of noteworthy corporations; it lays bare the close and uneasy relationship between the state and private business, particularly as companies expand beyond their borders. In the South African context, it also reveals the further complicated role of business in transforming the state. By focusing on archive material, company reports, academic text and government text, the book charts the making of the political economy of mobile telephony in South Africa and its implications for its foreign policy.

Telecommunications represents a critical component of the global political economy. The capacity to communicate over vast distances and control this technology is a determining factor in which nations actively participate in the international economy and remain on the periphery. Mobile telephony catalysed a revolution in telecommunications, facilitating individual access to telephony and ensuring near-universal internet connectivity. For African nations, mobile telephony presents a unique opportunity to bridge the digital divide and accelerate economic development. Examining the context around telecommunications illuminates the embedded nature of private sector firms. The fact that large telecommunication firms can influence statecraft and governance of a critical economic sector has far-reaching implications.

Derrick Cogburn points out that South Africa's Information and Communications Technology (ICT) development has 'paralleled and at times exceeded' the developments in the rest of the developed world. He points out that South Africa had access to commercial telegraphy less than two years after the United States. Furthermore, the South African Department of Post and Telecommunications was one of the first in the world to convert to digital switching in 1978 (Cogburn, 1998: 5). Even the proliferation of mobile telephony was not too far off from developed countries. In the United States, in the 1940s, mobile telephony was available to a limited extent in automobiles. In the 1970s, mobile handsets came into use.

Although South Africa only came onto the scene with commercial mobile telephony in the 1990s, it is fast catching up with the rest of the developed world. It is on the cutting edge of technology and innovation. The book focuses on the political economy of mobile telephony in South Africa. It sheds light on the interplay of power, values and political dynamics that colour the creation and arguably the success of South African-born mobile telephony giants MTN and Vodacom.

Few books consider the political economy of ICTs in an African context. One of the few that fill this important gap is Iginio Gagliardone's (2016) *The Politics of Technology in Africa*. Using Ethiopia as a case study, Gagliardone makes the case that ICTs in Africa are employed to serve political goals and are, therefore, not neutral actors. In this book I complement Gagliardone's work by further expanding the discussions around the political economy of ICTs in Africa. I home in on South African mobile telephony, specifically the network operators MTN and Vodacom. These companies have been the primary agents for the expansion of ICTs on the African continent, and their internal and external growth reflects South Africa's political and economic transformation.

In this book, I also speak to Liesel Louw-Vaudran's (2016) title *Superpower or Neocolonialist? South Africa in Africa*. Louw-Vaudren considers South Africa's corporate expansion on the African continent by the state and the private sector. She looks at the major corporate players such as MTN,

Vodacom and Standard Bank. She considers South Africa's image on the rest of the continent and if it could maintain its 'apparent' leadership position. My approach is dissimilar to Louw-Vaudren's; even though I consider South Africa's corporate expansion, my central concern is the internal and external dynamics between South African mobile telephony companies and the state.

Additionally, I dissect South Africa's private sector structure, offering a template for understanding broader patterns across the continent. Antoinette Handley (2008) conducts a similar inquiry in her work *Business and the State in Africa: Economic Policy-Making in the Neo-liberal Era*. Handley astutely illuminates the relationships between the South African state and the business community. She chronicles the evolution of the private sector, mindful of its historical backdrop, and elucidates how and why companies constructively engage in economic policy discourse. Handley adopts a broader perspective, while I focus on mobile telephony to provide a fresh perspective on the interplay between state and business in South Africa's political economic landscape and the implications for its corporate expansion beyond its borders.

MNCs and the political economy: A theoretical discussion

Theorists from the neo-Gramscian and Marxist schools of thought have extensively studied the role of multinational corporations (MNCs) as critical players in the global political economy. What methods can a state employ to exert influence over other states? According to Russett, influence can be cultivated by establishing a community characterized by shared kinship, loyalties, values and a collective sense of belonging (Russett, 1974: 259). This strategy is often referred to as a 'smart power' approach in the literature on international political economy. It involves significant investment in alliances, partnerships and institutions but also requires what is known as 'contextual intelligence'. Matthew Kutz offers a comprehensive definition of this concept:

> The ability to quickly and intuitively recognise and diagnose the dynamic contextual variables inherent in an event or circumstance and results in intentional adjustment of behavior in order to exert appropriate influence in that context. (Kutz, 2008)

MNCs, especially network operators, can be considered influential players in the political economy. The concept of the transnational capital class (TCC) provides an intriguing perspective on the role of MNCs in the political economy, portraying them as a dominant force. This notion aids in understanding the various ways MNCs wield their power within the global system and how this can be extrapolated to a regional political economy.

Sklair (2000, 2002) suggests that the TCC has gained increased control over the global system through capitalist globalization, subtly influencing regional and global policies. A prime example is the European Roundtable of Industrialists (ERT), which represents how the TCC functions. The ERT played a key role in establishing the European Centre for Infrastructure, which convinced the European Commission to initiate the European Networks, a collection of 150 environmentally conscious infrastructure projects, without needing approval from the European Parliament (Sklair, 2002: 145–6).

On the other hand, Upendra Acharya (2013: 948–9) presents an alternative viewpoint, arguing for an alliance between corporations and states to enhance state influence and market access until corporations reach or surpass the power of their home states. In this book, I adopt a less conservative stance than Acharya but recognize the potential for misuse of such relationships.

The political economy of mobile telephony

Control of communication lines has always been a point of contestation. It has always been 'not only a business but also a political enterprise, the subject of great-power rivalries' (Headrick and Griset 2001: 543). During the age of imperialism, the great powers of the time, Britain, France and Germany, fought not only over dominance over territory and resources but also for supremacy in the bourgeoning telecommunications industry. This degree of contestation was not simply about capturing business interests but also about securing diplomatic communication lines and the possible influence on the course of a war. The proliferation of early telecommunications corporations shaped the future state-business nexus worldwide, including on the African continent. For instance, the Tiv people, common in the middle belt of Nigeria, fought British attempts at installing telegraph lines over their land in the 1900s. The Tiv were against the Royal Niger Company's installation of telegraph lines across their agricultural land (Dorward, 1969; Ikimẹ, 1973). The lines represented a way for the British to connect to the Crown while marking future territory. Another example from Southern Africa was during the Anglo-Boer War in South Africa (1899–1990), portable phones provided by Swedish company Ericsson gave the British an upper hand in the war (Wickman, n.d; Austin, 2001). The introduction of new technology, ostensibly by a private company, changed the trajectory of a pivotal conflict. Interestingly, Ericsson continued to have favoured access through several South African administrations – even being one of MTN's first hardware suppliers decades later. Throughout history, we see how technology and business can alter political and social outcomes. Speaking on the creation of the telegraph, James Carey describes it as

> the first communications empire and the prototype of the many industrial empires that were to follow. The telegraph, in conjunction with the railroad, provided the setting in which modern techniques for the management of complex enterprises were first worked out, though for the telegraph in what was eventually monopolistic circumstances' (Carey 2009: 2).

In the late twentieth century and onto the twenty-first century, telecommunications still has import for the African continent regarding business, governance and diplomacy. South African companies, MTN and Vodacom, spearheaded the introduction of mobile telephony to Africa in the 1990s. The birth of two telephony giants, MTN and Vodacom, paralleled the ushering in of South Africa's democratic dispensation. As South Africa metamorphosed from a 'pariah state' to the bastion of 'moral capital', burgeoning industrialists and politicians saw abundant opportunities beyond the Limpopo. Less than twenty years later, South African mobile telephony companies came to dominate the African telecommunications space from Lagos to Khartoum.

The internet and mobile technology have become potent instruments for political mobilization and social transformation in Africa. This is evidenced by the numerous instances of governments restricting access to these services during periods of crisis. During Mozambique's 2010 'Bread Riots,' protesters leveraged short message service (SMS) technology to coordinate a week-long strike against rising prices of essential goods, ultimately compelling the government to reverse its decision (Jacobs and Duarte, 2010; BBC, 2010).

In 2016, eleven African nations imposed internet blackouts, including Zimbabwe, which blocked access to WhatsApp and Twitter amidst protests against the state (Shezi, 2017). In 2017 the Cameroonian government severed internet connectivity in the country's English-speaking regions (BBC News, 2017). In 2021 and 2022, there were approximately fifteen internet shut downs each year in response to civil tension (Access Now, 2022, 2023). The fact that many governments have responded to such challenges by ordering mobile network operators (MNOs) to shut down their services points to the power of mobile telephony, especially on a continent where more people have access to mobile phones than to electricity or piped water (Parke, 2016). According to an Afrobarometer infrastructure report, in the thirty-five countries surveyed, 93 per cent of citizens had access to a mobile phone, but only 65 per cent had access to electricity (Wambua and Balongo, 2016). This gives MNOs an incredible amount of structural power, as their services are widely used and can facilitate both development and dissent. However, it also raises questions about who controls the technology and how it affects the political economy of mobile telephony and has implications for the countries where the MNOs operate.

Chapter outline

The book is divided into three thematic structures. Chapters 1 and 2 are dedicated to concept building and exploring the relationship between technology and African international relations. Chapters 3 and 4 examine South Africa's corporate expansion on the African continent and the embedding of mobile telephony companies in South Africa's foreign policy. Chapters 5 to 7 focus on the evolution of mobile telephony companies in Africa and conclude the analysis.

1

Mobile telephony in South Africa

Mobile telephony has profoundly impacted the African continent since its introduction in the 1990s; widespread mobile telephony proliferation began around 1992. However, Mauritius, South Africa and a handful of North African states had access to mobile telephony before that. Despite its relatively late arrival compared to the rest of the world, the growth of mobile teledensity has been nothing short of miraculous. However, the benefits of mobile telephony extend far beyond mere connectivity. Access to markets and resources has been dramatically simplified, and there is a strong correlation between economic growth and telecommunication penetration, mainly through mobile telephony (Aker and Mbiti, 2010; Haftu, 2019).

ICT infrastructure across the continent has meant that mobile telephony has served as a bridge, providing African states with an opportunity to 'leapfrog' technological developmental stages (Evans, 2018).

Telecommunications, particularly mobile telephony, has become intricately woven into the social, political and economic fabric of much of the urban life of African nations. Mobile telephony is used in many domains, from social inclusion to financial inclusion through services such as mobile money. For instance, the emergence of mobile financial services, exemplified by the Kenyan mobile money service, M-PESA, has sprung up across the continent and offers unprecedented access to banking services for individuals traditionally excluded from these systems. Mobile money transactions in Africa account for 70 per cent of the world's transactions (Onyango, 2022).

In the book *Digital Democracy, Analogue Politics*, Nanjala Nyabola (2018) illustrates how access to the internet has changed politics in Kenya. Digital platforms have affected everything from elections to civic participation. Undoubtedly, to differing degrees, the internet has reshaped the fabric of society – particularly in urban contexts. However, as Nyabola asserts, 'No amount of digital innovation can paper over analogue failures, deliberate or otherwise' (Nyabola, 2018: 13). Even though Nyabola asserts in terms of democratic participation, I include unsuccessful attempts to ensure that the basic infrastructure and systems are in place to harness the digital age properly are also an analogue failure. My assertion applies to everything from the economy to social inclusion.

Meaningful participation in the global digital ecosystem – and by extension the digital economy – is vital to Africa's future. This vision is encapsulated in *Agenda 2063*, the plan for the continent's development, and the *African Union's Digital Transformation Strategy for 2020*. The critical elements of success (or crisis) are there. Most importantly, the region has a youthful demographic; by 2030, it's projected that over 42 per cent of the world's youth will be African (El Habiti, 2022). The failure to harness and correctly direct this demographic between ages fifteen to twenty-nine would be a missed opportunity for African states. This generation is most likely to have mobile access even though digital literacy is an area that still needs to be improved (DO4Africa, 2022; Statista, 2023).

Mobile broadband drives much of Africa's internet access, so many more people subscribe for access. Yet the cost of the devices and connections surpasses the rest of the world. These growth patterns and inconsistencies are not surprising. Minges and Kelly (1994: 13) observed three paradoxes that beset Africa's telecommunications sector in the early 1990s.

First, the level of telecommunication traffic per inhabitant (under 1 minute) contrasts with the fact that, as a whole, Africa had the highest international telecommunications traffic per subscriber (213 minutes), pointing to supply issues. Second, although Africans had the lowest revenue levels per inhabitant globally, telecommunications operators in Africa had the highest pre-tax profitability. Minges and Kelly surmise that the disparity might be because only the wealthiest segment of the population and sectors of the economy that are foreign-owned and focused on exports have access to telecommunications services. Third, the cost of installing a telephone line in Africa is the highest globally, even though the average income per consumer is the lowest. They point out that if the costs were 'closer to the world average, the region would add four times as many lines a year' (Ibid.).

Mobile technology has enabled African countries to close the telecommunications coverage gap as mobile towers are cheaper to deploy than fixed-line infrastructure. The paradox that characterized the 1980s and 1990s persists, highlighting the ongoing structural challenges in Africa. This situation is paradoxical in itself; while the issues with fixed-line telephony paved the way for the success of mobile telephony, the costs remain remarkably high.

Mobile voice packages and mobile data remain the most expensive in the world, with Africans paying approximately '200 per cent' of the global average for voice calls (Kazeem, 2020). Moreover, the average cost of individual data per gigabyte (GB) significantly exceeds the gross national product (GDP) of many African countries. For instance, Equatorial Guinea has Africa's highest mobile data cost at $49.67 per GB. Sao Tome and Principe, Malawi, Chad and Namibia follow, with prices ranging from $22.37 to $30.97 per GB (Oluwole, 2023). However, these countries have relatively low GDPs ranging from $0.55 billion to $13.6 billion (World Bank/OECD, 2023). The costs can be attributed to challenges in infrastructure development, regulatory hurdles, substantial telecom taxes and difficulties in establishing infrastructure in rural regions(Deloitte, 2014: 6–15; Oluwole, 2023).

Despite these challenges, mobile technology remains the cornerstone of Africa's digital future. It has not only propelled the telecommunications industry in Africa into the twenty-first century, but it is also significantly bridging the digital gap.

In 2021, mobile services in Africa (excluding North Africa) gained a subscriber base of 515 million, equating to 46 per cent of the region's population; this figure marks an increase of nearly twenty million compared to the previous year, 2020 (GSMA, 2022: 4). These growth rates make Africa one of the fastest growing regions in the world.

By the end of 2022, 214 mobile operators rolled out commercial 5G services in eighty-one markets worldwide; the technology has become dominant in the markets, leading to 5G adoption. For example, in South Korea, almost half of the mobile connections are 5G, and more than a third in the United States (Ibid.: 14). However, less than five African countries had commercial 5G services relying instead on legacy 2G and 3G networks. At the end of 2021, only 17 per cent of the total mobile connections in sub-Saharan Africa were 4G, compared to the global average of 59 per cent (Ibid.). For African countries, migrating from legacy services equates to better participation in the digital economy.

Outside of popular accounts of the *mobile miracle*, the story of mobile telephony in South Africa reflects a country and a continent in transition. Amidst the shifts in the African telephony landscape precipitated by epoch moments such as the Cold War, South African mobile network operators became familiar across the African continent. The dearth of communications infrastructure across the continent, the spread of neoliberalism and the hunger for *home-grown* solutions to Africa's problems created an environment ripe for the development of corporate giants. This growth represents both a boon and a burden for the South African government. On the one hand, the public image, and consequently the ability to project power, puts South Africa in a favourable position. On the other hand, South Africa's regulatory environment is ill-equipped to deal with the complex operations of external entities outside its territorial borders.

As South Africa became democratically inclusive, it became a force with significant moral capital. South Africa capitalized on this and positioned itself in a place of prominence. Upon recognizing that its future was inextricably linked to the continent, it pushed for regional integration and the renewal of Africa's fortunes under the banner of the 'African Renaissance' and the New Partnership for Africa's Development (NEPAD) programme. Daniel, Naidoo and Naidu argue that South African firms were well placed to take advantage of the outdated infrastructure on the continent in the 1990s as global interests at the time were focused on the Eastern European market (Daniel, Naidu and Naidoo, 2003). With time, it became increasingly apparent that South African companies would be the economic hands and feet of this integration drive because they had access to capital and technology.

These recent African mobile connectivity gains do not provide a complete picture of the inherent challenges of providing mobile telephony, including universal Internet provision; as Mark Warschauer suggests, 'While technology can shape social relations, social relations shape how technology is developed and deployed'(Warschauer, 2003: 301). The implications and opportunities were unparalleled; with the weakening of state telecom monopolies and the explosion in the mobile markets, the South African industry has been uniquely placed to take advantage of these opportunities, given the country's formidable technological infrastructure, access to capital and human resources. Waverman, Meschi, and Fuss (2005: 11) point out that mobile phones are playing the same role that fixed lines 'played in the richer economies in the 1970s and 1980s … countries with under-developed fixed-line networks have achieved rapid mobile telephony growth with much less investment than fixed-line networks would have needed'.

From a political economy perspective, the entity – be it a state or corporation – that spearheads the current wave of evolution of the telecommunications sector has the power to shape the existing system. This means that by leading the telecommunications revolution through mobile telephony, the country at the forefront can engage other nations in multilateral agreements, reinforcing its leadership position. Furthermore, states with advanced technology assert dominance, as evidenced by historical hegemons such as the United Kingdom, the Netherlands and the United States.

South Africa, owing to its history and developments in the international political economy, was fertile ground for South African corporations in the mobile telephony space to dominate the industry and expand on the continent rapidly.

A story of transition and transformation

The introduction of telecommunications technology, specifically telegraphy, to Africa in the 1870s was facilitated by colonial powers. Its strong European linkages enabled South Africa's early adoption of communication technologies, which provided the technical expertise necessary for effective implementation and utilization. In 1891, just a few years after the invention of the telephone in 1878, South Africa established its first telephony network to complement its existing telegraph network. South Africa's active engagement in the telegraphy, telephony and radio sectors can be attributed to its unique position as an extension of the British Empire and the presence of technical expertise within the country.

In the 1870s, the British South Africa Company (BSAC), founded by Cecil John Rhodes, constructed telegraph and railway lines primarily in Southern Africa. These lines linked African colonies and facilitated communication among colonial officials (Dhliwayo, 2013). The first iterations of Rhodes's dream of connecting Africa from *Cape-to-Cairo* found expression in the development of these lines, which all but halted construction after he died (Great Britain Colonial Office, 1935: 46). Rhodes was so intent on his grand imperial project that he formed the African Transcontinental Telegraph Company (ATTC); he visited the German emperor, William, in 1899, to broker a deal about building more telegraph lines on German territories in Africa (*Brisbane Courier*, 1899). However, intra-African telecommunications remained problematic even after the end of colonialism. The African Postal and Telecommunications Union (APTU), founded in 1935, attempted to address this issue but had limited reach, covering only southern Africa and being funded and managed from Pretoria (Mays, 2015). The telegraph lines are emblematic of the intersection of commerce, the state and international politics in communications infrastructure and deployment. Any story of communications or telephony in South Africa is replete without acknowledging the colonial logic behind the infrastructure deployment and how these logics manifests in every facet of telephony. Rhodes not only set out the blueprint to South Africa's telecommunications services and the vision for its reach, but he also set out what is arguably the outline for apartheid and an extension of racial hierarchization that underpinned colonial expansion. The Glen Grey Act was enacted in 1894 by the parliament of the Cape Colony while Rhodes was the prime minister. Rhodes's so-called Bill for Africa helped enforce the segregation of native Africans and limit their economic options (Bouch, 1993; Thompson and Nicholls, 1993).

During apartheid, South Africa's telecommunications services were highly segregated (Horwitz, 1999: 36, 76), with minimal access for the Black African population. Despite this, the republic had leaders and technocrats who recognized the importance and potential of telecommunications. One such individual was Hendrik Johannes van der Bijl, Eskom's first chief executive officer. Van der Bijl was an engineer and administrator with numerous patents, including creating the thermionic valve, an essential component for wireless telegraphy (Vermeulen, 1998, 2454). The administration ensured South Africa remained active in shaping the regional and international telecommunications regulatory framework while improving its white population's telecommunications access.

Table 1 Racial Teledensity 1990–2000

Access Type	Percentage of Households
Overall	81.6
African/Black	75.4
Coloured	94.3
Indian/Asian	98.7
White	99.1

Source: Adapted from Gillwald (2002: 5).

South African Posts and Telecommunications (SAPT) controlled telecommunications and postal services. Under the Union of South Africa, SAPT was overseen by the British Treasury and its revenue and expenditure were subject to parliamentary budgetary cycles. By the time South Africa became a republic, the SAPT's operations reflected the state's political ideology, its teledensity and infrastructure development were unequal and discriminatory to Black Africans (Horwitz, 2001a: 76; see Table 1).

By the late 1980s, South Africa was amidst a debt crisis. As a result, it had no choice but to relinquish a level of control of its state-owned enterprises. South Africa's external debt had reached over 66 per cent of its GDP by the time President Botha had made his infamous 'crossing the Rubicon speech', which caused the South African markets to crash further in 1985 as it appeared that the apartheid government was digging in its heels about the transition to democracy and the releasing of Nelson Mandela after declaring a state of emergency earlier in the year (Frankel, 1985; Kahn, 1991: 66–7; Laurence, 2018). The road was open to a sweeping transformation of every sector despite initial resistance to structural economic reform; the South African government had no choice but to relook at its economic structural model (Grant, 1985). Notably, the investigation into the viability of the South African parastatals, conducted by W. J. de Villiers (known as the De Villiers Report), was instrumental in the ultimate liberalization of state-owned enterprises, including telecommunication. As Horwitz reminds us, loosening the reins intersected with the transition to democracy. 'Whether or not government intended it, the effect of such privati[s]ation would have been to take the parastatals out of the hands of the newly empowered black majority in a democratic dispensation' (Horwitz, 1999: 211).

The African National Congress (ANC) went back and forth on whether or not they would privatize, with resistance coming from the Black workers union, the Postal and Telecommunications Workers Association (POTWA) being adamantly against it. Commercialization was put forward as a middle-ground alternative. Privatization of the sector began in the dying days of apartheid in 1989 and was completed by the ANC government. Internal and external factors influenced the decision to continue with the privatization process. The most important of the external factors was that the telecommunications sector globally was liberalizing and leaning towards privatization. Internally, the introduction of mobile telephony and improving sector efficiency made privatization attractive despite resistance.

State-business relations and parastatals

The transition period between the apartheid state and the incoming ANC government was a dance between priorities. There was a need to rescue the state from the financial quagmire it found itself in, while structurally transforming the economy to ensure that the marginalized Black majority were meaningfully included. The relationship between the state, private businesses and public corporations has always been fraught with tensions, even more so during the transition period. State-owned enterprises (SOEs), such as Eskom and Telkom, operated different logic and mandate, and even during the supposed move towards commercialization, they remained instruments of the State for providing public goods or stimulating economic development (Cuervo-Cazurra et al., 2014). As Bowman underscores, the parastatals primary role is 'lowering strategic business costs by providing cheap access to network infrastructure' (Bowman, 2020: 400). In the case of telecommunications, South Africa's telecommunications was one of the most sophisticated on the continent at the time. Moreover, South Africa was able to manufacture telecommunications components; however, with time, due to sanctions, the state could not continue that role and became a middle party between larger corporations and the general customer.

Importantly, SOEs were a bastion for white labour. They were instrumental 'to the formation of an Afrikaner middle class through employment and through preferential procurement' (Ibid.: 404). To understand the reasons behind this, we must consider the nature of racial and economic hierarchy in South Africa in the early 1900s. Although the white minority population were at the top of the hierarchy, white Afrikaners found themselves at the bottom of the economic ladder for a number of reasons (Sachs, 1957: 12–15; Fourie, 2006). Poverty was so endemic amongst Afrikaners in the 1900s that the Carnegie Corporation commissioned a study on poverty amongst whites in 1932. The report's findings further emboldened the development of segregationist policies and supported racial segregation (Seekings 2008). As Sampie Terreblanche notes:

> The real embourgeoisment of the Afrikaners took place from 1948 to 1974. In implementing its policy of 'affirmative action' or social upliftment from 1948, the NP enacted additional repressive and discriminatory legislation ... The public sector and public corporations were also used to create lucrative job opportunities for Afrikaners. All these redistributive policies significantly improved the economic position of Afrikaners. An outstanding feature of the NP policy was its rather interventionist character. The NP intervened in the economy in all sorts of ways and used public corporations as a vehicle for Afrikaner upliftment. (Terreblanche, 2005: 5)

Any analysis of why the New ANC was married to the idea of affirmative action and, ultimately, a programme of Black economic empowerment is because they saw that it worked with the upliftment of Afrikaners. To make these policies work, the state must still follow an interventionist stance despite external pressures to liberalize and privatize. This was acutely obvious in the telecommunications sector. As Alan Levy et al. (2021: 2) point out, South Africa's political settlement that led the way to democracy was undergirded by four 'sub bargains':

1. An economic deal between the primarily white economic elite and the ANC ruling party. The deal included assurances of private property protection, respect for the rule of law and transformation through Black economic empowerment (BEE);
2. An embedded undertaking within the ANC, and all its structures, that the transformed economy would be channelled towards 'a shared national purpose';

3. The creation and protection of the new predominantly Black middle class through education, job creation and ending racial discrimination; and
4. The reduction of extreme poverty through a redistribution of public resources.

The decision to pursue privatization post-apartheid was undoubtedly complicated by politics and yet necessary to transforming the economy, building a Black middle class and shaping 'a bond between the state and capital' (Pitcher, 2012: 243). A triangular relationship exists between the state, capital (business) and the middle class. The growth or stability of the middle class often signifies the success of the economic and social policies of the State. In the South African context, the existence and growth of the Black middle class illustrate that the ANC-led government can change the economy's structure to make it more racially inclusive (Burger et al, 2013; Melber, 2017; Korhonen, 2018). Before 1994, the state's wealth primarily rested amongst the minority white population. Whether there has been a real structural change to the economy is a matter of debate. The state institutes policies that affect the growth of capital and the well-being of the middle class. And capital is responsible for economic growth, and ultimately the social and economic stability of the State.

But privatization could not be done wholesale because of the strategic importance of state assets. These sub-bargains affected all parts of the economy, including telecommunications. Similarly, the shift towards commercialization or liberalization of telecommunications occurred amidst underlying conflicts within the ANC alliance regarding the economic direction of the new South Africa.

In subsequent chapters, I illustrate how mobile telephony helped forge a Black empowerment business model that linked up the state, capital and the bourgeoning Black elite. Also, how the state-business nexus shapes South Africa's foreign policy in Africa and boosts its soft power projection. Yet, on the other hand, these new private firms still have to balance their profit motives with the state's expectations and incentives and handle the corresponding conflicts. Although my main argument is that as private actors, South African MNCs contribute to consolidating South Africa's leadership role and aligning its economic interests with its foreign policy goals, it still raises the question of who is driving the agenda and setting the terms of engagement in Africa. There is concern among the ANC alliance partners and other African nations that the expansion into Africa is being led and influenced by businesses which are predominantly white and foreign-influenced, rather than being guided by South Africa's stated foreign policy orientation. This has raised fears of potential exploitation or the establishment of a malignant hegemony (as was seen during apartheid) if these expansion efforts do not align with the state's African and Global South centred ideology.

Post-apartheid South Africa's various economic programs, such as the Reconstruction and Development Programme (RDP), the Growth, Employment and Redistribution (GEAR) plan, the Accelerated Shared Growth Initiative of South Africa (AsgiSA) and the National Development Plan (NDP) 2030, had various degrees of success and drivers. The RDP base document, in particular, underscored telecommunications' importance to South Africa's transformation. It states that:

> 2.8.3 The telecommunications sector is an indispensable backbone for the development of all other socio-economic sectors. An effective telecommunications infrastructure which includes universal access is essential to enable the delivery of basic services and the reconstruction and development of deprived areas. ('The O'Malley Archives', 2024)

Throughout the base document, and as reflected in the final document that was implemented, there was recognition that there needs to be a way to provide for the most marginalized in society. The RDP failed to achieve consensus within the tripartite alliance due to opposition from those who favoured a neoliberal approach, such as the business community, despite public pronouncements to the contrary (Edigheji, 2003: 93).

The state-business relationship (SBR) in South Africa is shaped by multiple factors, including race, intra-party ideological differences and the political clout of organized labour. The Congress of South African Trade Unions (COSATU), as a member of the tripartite alliance with the ANC and the South African Communist Party (SACP), has the ability to influence policy decisions at the party level. A case in point is suspending a controversial highway-tolling project in Gauteng due to pressure from COSATU. The National Economic Development and Labour Council (NEDLAC) is the primary institutional mechanism for coordinating SBR. However, some analysts argue that the State has undermined NEDLAC's effectiveness by engaging with businesses and labour on policy issues outside its purview (te Velde and Leftwich, 2010: 18; Ensor 2015).

> Organised business, in turn, had little incentive to engage with the State over its macro-economic management, and became increasingly sceptical of Nedlac as a forum for dealing with other policy issues. Collective action by business was undermined further by the fact that the established corporations and the emerging 'empowered' black business elite could pursue their individual agendas more successfully through direct engagement with government. (Nattrass and Seekings 2010: 4)

The NEDLAC was formed in 1995 as a formal institution for government to engage with organized labour, organized business and civil society on issues related to socio-economic policy (NEDLAC online). More specifically, its constituencies are:

- The Government: Departments of Labour; Treasury; Public Works; and, Trade and Industry.
- Organized Labour: Congress of South African Trade Unions (COSATU), Federation of Unions in South Africa (FEDUSA) and the National Council of Trade Unions (NACTU).
- Organized Business: Business Unity South Africa (BUSA) (MTN South Africa is a member of BUSA)
- Civil society groupings such as the Women's National Coalition, the South African National Civics Organisation, the South African Youth Council, Disabled People of South Africa, The South African National Apex Cooperatives, and the Financial Sector Campaign Coalition.

Lauded as an example of cooperative governance, Edigheji highlights that the precursors to NEDLAC were the National Economic Forum (NEF) and the restructuring of the National Manpower Commission (NMC). They were the ways that COSATU hampered the ability of the failing apartheid regime from, 'unilaterally imposing social and economic policy' (Edigheji 2003: 77). For instance, through the pressure of a wave of collective industrial action in the late 1980s, COSATU managed to secure a place on the NMC (which was initially a 'whites' only labour formation). The purpose of the NMC was to advise the State on labour affairs (Bendix, 2010: 244). Soon after their inclusion in 1991, there was a push by both organized business and labour to have input in economic reform which saw the birth of the NEF after a devastating two-day general strike in the manufacturing sector called by COSATU (Houston, Libenberg and Dichaba 2001: 35). The strike began as a protest against value-added tax. Still, forming the NEF became a key demand (Ibid.). COSATU envisioned the NEF as a 'bargaining forum' that parliament would be compelled to ratify (Skålnes, 1997: 163). Edigheji notes that 'the origin of consultative decision-making in the South African context was not about strengthening governance capacity but about reducing State capacity' (2003: 77). This discussion is important as it goes to the heart of SBR and corporate expansion. It demonstrates that the State and organized business relationship is as complicated under the ANC-led government as it was under the old regime. However, despite these complications and the failure to create a corporatist arrangement, the interests of the private sector, as is seen with Treasury and Department of International Relations and Cooperation (DIRCO) policy initiatives, is still embedded in South Africa's statecraft. A possible reason for this is that the State recognizes that the corporate sector can still be used as a tool for the State, particularly for meeting its development agenda. In the case of MTN, as well as Vodacom and Cell C[1] later on, their licensing is linked to providing services to the community's marginalized sectors.

Arguably, the State rendered NEDLAC toothless by sidestepping it in the key area of macroeconomic strategy, evidenced by introducing the Growth, Employment and Redistribution strategy (GEAR) in 1996 (see Figure 1). GEAR can be characterized as a self-imposed structural adjustment program. It aimed to promote socio-economic development, similar to the 1994 RDP. However, GEAR's ideological approach differed significantly from the RDP. While the RDP aimed to address socio-economic challenges by prioritizing basic social needs such as housing, education, services, health and land, GEAR adopted a more technocratic approach that was less socialist-leaning and more appealing to the business community. This resulted in a focus on reducing

[1] Co-founded in November 2001 by Lehlohonolo (Lambert) Moloi, a former South African military commander and a former commander of the ANC's military wing, Umkhonto weSizwe. Cell C is South Africa's third MNO, and its structuring is emblematic of the BEE arrangements for MNOs – as seen with MTN and Vodacom. Its complex shareholding includes CellSaf as its empowerment partner and Saudi based Ogar Telecom as the main shareholder (see TeleGeography, 2016; Mochiko, 2017).

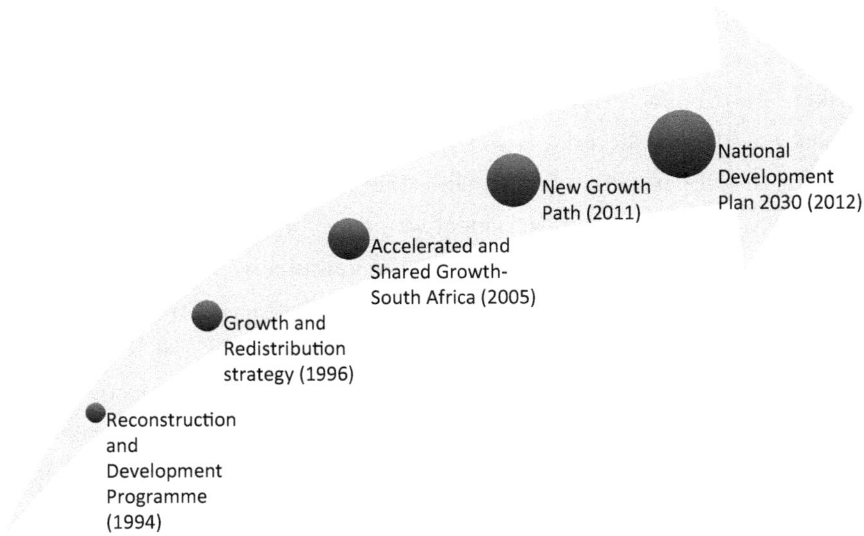

Figure 1 Key South African macroeconomic strategies post-1994.

Source: Author's Graphic Representation.

government spending and improving conditions for international trade, such as reducing import controls.

The AsgiSA was introduced as a strategic framework to address the economy's weaknesses and promote inclusive growth through increased government investment in infrastructure. Priority sectors included business process outsourcing and tourism. The main objectives of AsgiSA were to boost infrastructure investment, target sectors with growth potential, develop skills, support small businesses, improve public administration and foster a conducive macroeconomic environment.

In 2010, the New Growth Path (NGP) was launched to create five million jobs and reduce unemployment to 15 per cent by 2020. Job creation is at the heart of the NGP, along with the vision of building an integrated African economy and forging partnerships. The NGP represents a shift in policy focus towards addressing the pressing challenge of unemployment and promoting sustainable and inclusive growth.

Nevertheless, between 1996 and 2012, South Africa's economic growth programs were characterized by a neoliberal thrust that emphasized the free movement of capital legitimizing and amplifying corporate power, particularly among elites in crucial sectors such as mining, manufacturing, agriculture and telecommunications. This has led to a concentration of wealth and influence in these critical areas. The corporate sector played a crucial role in enabling the State to exert its influence across the continent and strengthening its capacity to negotiate a more favourable export, trading and investment framework. Expansion into Africa also improves South Africa's economic growth, including strengthening regional and global value chains, which ultimately brings jobs to South Africa (Stern and Ramkolowan, 2021).

This could be viewed as an apparent retreat from the interventionist stance that characterized the previous regime and ensured Telkom's continued monopoly. The move towards a new SBR was truly underway.

Conclusion

Telecommunications has historically played a pivotal role in the South African state apparatus, serving as a means of control and an indicator of the state's evolving priorities. This became particularly evident during the 1980s, as the National Party began to distance telecommunications from direct ministerial oversight, establishing a state-owned enterprise akin to a private corporation. However, it is essential to note that the apartheid government never entirely relinquished control over the telecommunications sector. In 1991, the SAPT was divided into two state-owned entities: Telkom and the Post Office. Telkom SA Limited was incorporated as a public company under the South African Companies Act, with the government transferring its telecommunications operations to Telkom. This decision was shaped by both the sociopolitical climate of the time and the recommendations of the De Villiers report (De Villiers and South Africa, 1989), which played a significant role in the restructuring and liberalization of the telecommunications industry.

As South Africa transitioned to democracy, its domestic political economy shifted from a capitalist model towards what could be characterized as an emerging corporatist model. Corporatism envisions the State orchestrating industrial relations between business and labour, acting as collective partners. Understanding South Africa's domestic economic and political model is essential for comprehending how multinational corporations in the telephony sector have gained considerable influence.

2

Reform of the telecommunications sector

The introduction of mobile telephony in South Africa occurred within the context of the economic and political transitions of the 1980s and early 1990s. Several competing interests were at play. The incumbent government was on its way out, and the country's financial outlook was poor. At the same time, their constituents, primarily white businesses, needed to secure their interests in the country. The incoming ANC government understood that they faced a mammoth task in taking over the reins and doing so from a developmental perspective. In other words, they had to ensure that most of the population, primarily Black and on the fringes of the economy, were catered for without alienating capital. At the same time, the ANC government was responsible for creating a more inclusive economy in the face of the incumbent's distinct push towards privatization. Due to its timing, the motives behind this push were questioned in many quarters. As Ben Fine and Zavareh Rustomjee (1996: 53, cited in Hentz, 2000: 205) maintain, 'The privatization program in South Africa was clearly motivated, at least in part, by the wish to remove state-owned assets beyond the control not of the state as such, but of those interests that were about to adopt state power.' It would be hard to disavow such contention in light of the shedding of many state assets, including parts of Telkom.

The creation of the state-owned enterprise Telkom in 1991 marked a critical turning point in the history of South Africa's telecommunications sector, as it shaped the emergence of the first mobile network operators (MNOs) and their relationship with the domestic regulator. The business model these MNOs developed in South Africa became the template for expanding beyond the country's borders. This is exemplified by (Mobile Telephone Network)

MTN, which initially operated its subsidiaries similarly, focusing on local empowerment in line with the African Renaissance ideology and NEPAD.

The 1994 RDP base document acknowledged the role of telecommunications in socio-economic development and the need for economic transformation through programmes such as BEE. Mobile telephony licences were introduced in this context, and the telecommunications regulatory regime was overhauled. The De Villiers report and the RDP base document identified a severe gap in service provision to mainly Black customers. Expanding fixed telephony infrastructure was costly, making mobile communication a viable alternative. The government decided to license two digital MNOs. The licensing of two digital MNOs was based on a report by Coopers and Lybrand, which recommended that Telkom be allowed to participate in one of the franchises, either independently or as part of a consortium, through a separate subsidiary (Horwitz, 1999: 217–20). However, the government released the tender without consultation or supporting legislation in April 1993, sparking controversy with the ANC, COSATU, and the Cellular Telephone Consultative Forum (Ibid.: 200). Ters Oosthuizen[1] was appointed to administer the process, drawing criticism from these stakeholders.

The introduction of mobile telephony was not well understood by either the ANC or the National Party (NP), according to industry insider Alan Knott-Craig (Knott-Craig, 2012). However, both parties saw the relatively low cost of mobile telephony as an opportunity for political gain. Horwitz cites an account of an NP government official seeking to reap the political benefits of mobile telephony before the ANC as a reason for proceeding with the licensing process outside of agreed-upon mechanisms (Horwitz, 1999: 221). The ANC opposed the issuing of mobile telephony licenses to private companies (Muller, 2014), likely due to concerns about job instability at Telkom resulting from competition. Mandela and F. W. De Klerk ultimately resolved their differences over various aspects of the mobile telephony regime, including whether to adopt the GSM standard through one-on-one discussions (Ibid.). Two mobile licenses were initially awarded: MTN Holdings and Vodacom.

[1] Ters Oosthuizen was the post-master general before the reform of telecommunications sector. He became the ICASA chairman. He was responsible for issuing MTN and Vodacom's operating licenses.

The battle for mobile licenses

The battle around the mobile telephony licenses illustrates the tension between the incumbent government, the ANC and its partners, and the market. It all comes back to the issues of regulation. Robert Horowitz points out that the Post Office Amendment Act of 1991 did not radically reform the sector as expected (see more on regulation reform in the next section); instead, it only deferred decisions around the telecommunications sector structure and regulations (Horwitz, 1999: 213). This was a problematic state of affairs as the country was changing, a new government was about to step into the position and the framework around one of the most critical sectors, arguably integral to the socio-economic sustainability and growth, of the state was essentially left unresolved. The effect, as Horwitz illustrates, was that the Department of Posts and Telecommunications (DTPS) was acting as regulator without the legal authoring to do so; legal amendment only replaced the DTPS with Telkom, essentially putting the newly formed company in the position of regulator and operator.

The real problem, as Horwitz points out, was the collapse of Convention for Democratic South Africa (CODESA).[2]

The result was that the state appeared to go ahead and make decisions for the sector without having an appropriate governance framework in place, and without buy-in from the ANC and the unions (Horwitz, 1999: 214).

The NP government went ahead and issued the telecommunications license without consultation; the people behind this decision were minister of transport and communications Welgemoed and Eugene van Rensburg, head of the Policy Unit on Public Enterprises of the Ministry of Mineral and Energy Affairs and Public Enterprises (Ibid.: 220).

> Another explanation of the government's actions put politics at center stage. According to one report, a senior government source was quoted as saying that the National Party government saw that it had the opportunity to provide telephones to the masses quickly through cellular service, and the Party had no intention of waiting and thereby permitting the ANC to reap

[2] CODESA was a series of multiparty negotiations between 1991 and 1992 that were used to chart the path for a democratic South Africa, including teasing out the interim constitution.

> the political benefits. Other observers invoked a political explanation of the more traditional kind, arguing that Welgemoed was a minister 'of the old school' – that is, of the imperious apartheid stripe – who would never allow 'interlopers' to make policy or interfere with his authority. (Ibid.)

Before the MTN and Vodacom licenses were issued, Telkom offered cellular services using first-generation technology, run on the c-450 network, and Siemens technology ad handsets. There were around 11,000 users when the first network was licensed (Business Day, 1993a: 8). What is clear is that the state had already a test run of offering mobile telephony, albeit via a small number of users (Business Day, 1993b: 3). This was arguably a precursor to Telkom's majority shareholding of the first license; the NP government appears to have intended to continue to direct the telecommunications sector via mobile telephony.

As Horwitz (1999: 222–3) aptly explains: The cellular tender process revealed the divergent interests and visions of the government, the ANC alliance, and the private sector. The government had already adopted the GSM standard[3] in principle, based on its long-standing relationship with Siemens, its main equipment supplier. Some praised the GSM system for its flexibility and potential to serve underprovided communities, but it was criticized by others for its high costs and complexity. On the other hand, the ANC alliance wanted cellular to be a separate public service offered by Telkom, in a way that would ensure universal access and rational planning. The ANC alliance accused the government of favouring private interests and foreign investors over the public good. It threatened to revoke the licenses if it came to power (see Sergeant, 1993b: 3; Susman Associate Press, 1993: 1). The NP government refused to halt the tender process, arguing it would undermine legal and economic stability. The conflict escalated in late 1993, with strikes, protests and high-level meetings between the two sides. In the midst of the struggle over cellular licenses, the South African government introduced a bill to amend the Post and Telecommunications Act. The ANC alliance interpreted this move as an attempt to deregulate Telkom and further undermine its vision of universal service. However, a compromise was eventually reached between the two sides.

[3] A European technology standard.

In exchange for the government's agreement to postpone the amendment bill, the ANC alliance withdrew its opposition to granting cellular licenses. This agreement temporarily defused the tensions between the government and the ANC alliance over telecommunications policy.

The precursor to radical regulatory reform

The White Paper of 1996, issued by the first democratic government, was a radical document that articulated three fundamental principles: first, it recognized telecommunications as a vital precondition for development; second, it initiated a process of liberalization in the sector; and third, it established a framework for economic empowerment of the historically marginalized Black population through equity, employment and entrepreneurship opportunities (The Ministry for Posts, Telecommunications and Broadcasting, 1996: chapter 4). This was one of the earliest expressions of the broader Black empowerment agenda that would shape the post-apartheid political economy.

Horwitz and Knott-Craig's analysis of the emergence of mobile telephony in South Africa suggests that the political economy model adopted in the early years of the democratic transition was influenced by a form of social corporatism, where the state sought to incorporate established business and labour interests into a new social pact. This pact aimed to transform the existing socio-economic order.

An essential criterion for the MNO tender application was that applicants had to demonstrate how their technology choices would enable high-volume and low-cost services, support local industry development and serve poor communities. The license also stipulated a quota of phones to be allocated to underserved areas (Horwitz Ibid.: 201).

All initial licenses for MNOs in South Africa included an empowerment component and a Universal Service Obligation (USO; see Table 2). A USO must legally provide network access to underserviced areas (e.g. rural areas and poor communities) at an affordable price – often below profitability (Intven and McCarthy, 2005). South Africa's USO requirements reflected global shifts in the regulatory environment for telephony. However, compared to global trends at the time, 'South Africa was relatively unusual in including

Table 2 Universal Service Obligations Imposed on Licensed Operators in South Africa

MNO	Coverage	Community Access Points
MTN	60% of the population within two years, 70% of the population within four years' timetable for specified coverage areas	500 community service telephones over five years - 1,250,000 SIM card connection packages - 125,000 handsets - 140 institutions for people with disabilities provided with internet access (ten terminals each) over three years - 5,000 public schools provided with internet access over eight years - (subject to approval of implementation plan)
Vodacom	60% of population within two years, 70% of population within four years' timetable for specified coverage areas	22,000 community service telephones in seventy specified areas over five years - 1,250,000 SIM card connection packages - 125,000 handsets - 140 institutions for people with disabilities provided with internet access (ten terminals each) over three years - 5,000 public schools provided with internet access over eight years - (subject to approval of implementation plan)
Cell-C	40% of area within one year (roaming) 8% of area within five years (own network) 80% of population within one year (roaming) 60% of population within five years (own network)	52,000 community service telephones in underserviced areas (with less than 10% fixed teledensity) - 1,250 000 SIM card connection packages – 125,000 handsets - 140 institutions for people with disabilities provided with internet access (ten terminals each) over three years - 5,000 public schools provided with internet access over eight years - (subject to approval of implementation plan)

Source: Adapted from Lewis (2013: 97).

USOs for its mobile operators', now a tactic recommended by the International Telecommunications Union (ITU) (Lewis, 2013: 96).

Along with the imposition of network geographic and population coverage requirements, the USO requirements for mobile focused on 'community service telephones' – essentially a public payphone on the mobile network 'freely accessible' to the 'general public'– with slightly differentiated requirements across the three licensees. Additional obligations were imposed on MTN and Vodacom in 2004 in return for access to additional spectrum. These were requirements to distribute free SIM cards and handsets, as well as provision of internet access to public schools and institutions for people with disabilities, and were later extended to Cell C (Ibid.).

To fully comprehend the complexities of the political economy of mobile telephony in South Africa, it is essential to understand the implications of BEE within the context of a liberalizing sector. The evolutionary trajectory of South Africa's MNOs encompasses addressing historical wrongs, adhering to basic market principles and responding to global changes.

Vodacom pioneered mobile telephony in South Africa in the early 1990s, and part of its vision was to provide services to the previously excluded and vulnerable segments of society who could benefit from enhanced communication capabilities for economic and social advancement. In the early 1990s, Vodacom's initial license terms were revised to include a provision that required Vodacom to supply '22,000 phones to previously underserved areas within 5 years' (Coetzer, 2008). One of the strategies that Vodacom employed to reach poor communities was to offer phones for rent within shops owned by local entrepreneurs. According to Carney (2008), in 2004, more than half of the traffic on Vodacom's network came from '4,400 entrepreneur-owned phone shops where customers rent access to phones by the minute'. The effect of these phone shops along with its first-mover innovation, was that Vodacom was quickly able to meet its license conditions and ensure that it dominated market share. An example of this was the transformative effect that community phone lines had in the eastern Cape (one of the least developed regions in South Africa. In 1996, in the Transkei,[4] only 19,000 phones were available to service an estimated population of four million people, with fixed-line infrastructure last being upgraded in 1988. Two base stations were subsequently built, resulting in over 10,000 calls a day made(Reck and Wood, 2003: 6). The community service rate, set by the regulator at R0. 5 per minute was below the regular service rate for prepaid calls, set at R2.75 per minute (Ibid.: 7).

Vodacom pioneered the provision of prepaid services in 1996 and prepaid internet access in 1998. This innovation expanded internet access

[4] The importance of connecting underserved locations cannot be understated. The Transkei, for instance, was a bantusan created for Xhosa speaking peoples in 1959. It became a so-called homeland under the apartheid system and was plagued by underdevelopment.

to individuals who could not qualify for contract services due to income constraints.

South African MNOs, particularly Vodacom and MTN, have had to navigate a complex regulatory, legal and diplomatic landscape when entering new markets. This includes South Africa's telecommunications regulations, transformation agenda, legal requirements for multinational corporations and emerging economic diplomacy framework. However, the relationship between these corporations and the state is not straightforward. While private companies are accountable to their shareholders and not bound by the state's developmental or transformational objectives, South African MNOs have initially closely aligned with the state's priorities. This reflects the historical engagement of South Africa's corporate sector with the state.

Regulatory evolution and South Africa's MNOs

Before the democratic transition, South Africa's telecommunication sector was under the direct control of the state. The DTPS was responsible for regulating and licensing the sector, without any independent oversight. This changed as the global trends of liberalization and privatization in the telecommunication sector influenced South Africa's policy reforms in the 1980s and 1990s. However, South Africa's reforms were not simply driven by economic logic, but also by the political context and the internal debates within the ANC alliance over the economic trajectory of the new democracy.

As discussed previously, the licensing of the MNOs was a contentious issue during the democratic transition negotiations. Vodacom was formed with Telkom as the majority shareholder, whereas MTN was formed as one of the first Black empowerment corporations. This was an untenable situation as the state was essentially regulating itself. By 1994, a telecommunications regulator was created: the South African Telecommunications Regulatory Authority (SATRA). SATRA was soon replaced due to South Africa's accession to the World Trade Organization's Fourth Protocol on Basic Telecommunications. Article 5 of the agreement's reference paper stated that an independent regulator had to be established. It stated that:

> the regulatory body is separate from, and not accountable to, any supplier of basic telecommunications services. The decisions of and the procedures used by regulators shall be impartial with respect to all market participants. (WTO, 1996)

In addition to South Africa's international obligations, there were still attempts by the incumbent to hold onto control of the sector. In telecommunications, this meant that the DTPS retained its authority under the Telecommunications Act of 1996, even though the act created provisions for regulating telecommunications and establishing independent regulators. This shows how political factors can affect economic policy and outcomes.

Several domestic initiatives aligned South Africa with its international obligations, including the National Telecommunication Forum, which led to establishing the Independent Communications Authority of South Africa (ICASA). This regulatory body was intended to reduce political interference in the telecommunications industry, but its council appointment process has undermined this objective. The discussion on regulation is crucial because it sheds light on where power resides domestically and highlights the current challenges and opportunities in South Africa's telecommunications sector.

Horwitz and Currie's interview with the former chief executive officer of ICASA, Nkateko Nyoka, highlighted that 'real power … lies in interpreting policy, a prerogative that rests with the Minister. And the Ministry is open – if only by virtue of its shareholding in Telkom – to industry (especially Telkom) lobbying' (Horwitz and Currie, 2007: 454). In 2002 the Electronic Communications and Transactions Act (ECT Act) of 2000 replaced the Telecommunications Act. Subsequently, in 2005, the Electronic Communications Act 36 replaced the ECT Act. However, chapter two of the act still gives the minister of communications the power to interpret and issue policy.

Mobile telephony is subject to stringent regulation (see Table 3) in all countries of operation; in South Africa, the regulatory authority is ICASA. It was among the first regulators on the continent to introduce number portability and reduce mobile interconnection rates. To improve the quality of service, ICASA is mandated to implement the Electronic Communications Amendment Act No 37, which aims to create a more favourable environment for investment and innovation. However, bureaucratic red tape delayed the

Table 3 Evolution of Telecommunications Regulation in South Africa

Year	Regulatory Authority	Description
Pre-1996	Department of Posts and Telecommunications and Telkom	The telecommunications industry in South Africa was subject to *central regulation*, overseen by the Department of Posts and Telecommunications – and later in conjunction with Telkom. Telkom held a monopoly on licensing and regulatory authority.
1993	Independent Broadcasting Authority (IBA) Act	The IBA Act instituted *an independent and unbiased regulatory body* to oversee the regulation of broadcast content and signal dissemination, representing a departure from the prior system of centralized regulation.
1996	South African Telecommunications Regulatory Authority (SATRA)	The *Telecommunications Act of 1996* instituted *SATRA* as the regulatory body for telecommunications, with the mandate to act in the public interest. This act effectively separated the functions of policy-making, regulation and implementation.
2000	Independent Communications Authority of South Africa (ICASA)	The ICASA Act of 2000 consolidated the IBA and SATRA into a single entity, ICASA, which assumed the role of the exclusive regulatory body for both broadcasting and telecommunications in South Africa.

Source: Compiled by author.

implementation of the act, resulting in operators facing difficulties in obtaining timely approval for infrastructure rollout.

In addition to frequent changes in leadership, the ministry responsible for telecommunications has been plagued by inefficiencies, including a lack of policy on spectrum allocation. Spectrum allocation is critical because radio frequency bands are required for various applications, including television and mobile phones. Mobile operators need sufficient spectrum to provide high-quality mobile broadband services with high data transfer rates and indoor and outdoor coverage. The transition from terrestrial to digital television was expected to free up additional spectrum, but South Africa missed the ITU deadline of June 2015 for this migration.

These dynamics are significant because they impact the performance and prospects of South African MNCs expanding their presence on the continent.

Despite this confusion, through its foreign policy, South Africa has demonstrated its understanding that its well-being is inextricably linked to that of the rest of the continent – including the telecommunications sector.

MNO multinationals have aligned their expansion strategies with South Africa's foreign policy objectives without an official state mandate. For instance, in a seminar presentation to parliament's International Relations Committee in 2007, former group executive of corporate affairs Mthobi Tyamzashe linked Vodacom's African operations to one of the pillars of South Africa's foreign policy – African prosperity and stability (Tyamzashe, 2007). Tyamzashe's presentation also included investment criteria that went beyond economic benefit for shareholders, such as 'government invitation' and existing 'bilateral trade agreements between South Africa and targeted countries' (Ibid.). Similarly, in an interview with MTN's Chris Maroleng, he described MTN's market choice as one that 'goes beyond pure commercial benefit defined in terms of profit maximization' (Maroleng, 2016).

This raises the question: why would MNO multinationals link their African expansion to South Africa's post-1994 foreign policy if there was no explicit government mandate for business? To answer this question, one needs to understand the evolution of South Africa's domestic telecommunications landscape and the critical actors involved in the pan-African expansion. The following two chapters will examine the MTN and Vodacom cases in more detail, but this subsection provides an overview of the issues at stake.

Conclusion

A few decades later, the South African government is at risk of losing the benefits it has accrued from mobile telephony, perhaps unwittingly. When mobile telephony was first introduced, the state had a significant stake in MTN and Vodacom (as discussed in more detail in subsequent chapters). It gradually sold its shares in these companies, which reduced the state's influence on the growth trajectory of MNOs. A fundamental problem, it can be argued, is the lack of a strategic vision that guides policy decision-making. This creates a vacuum that can be filled by the agendas of external actors. In other words, without a clear plan, the government has allowed others to set the agenda for ICT and mobile telephony.

South Africa has been a leading ICT player due to its technological expertise and developed mobile industry. Dagada notes that several innovations in mobile telephony were 'invented or refined in South Africa', such as prepaid services, mobile fax transmissions and mobile banking (Dagada, 2012: 53). However, South Africa has started to lag behind countries such as Kenya and Nigeria, which are both investing heavily in upgrading their infrastructure. Kenya's successful integration of ICT into the private and public sectors is well documented (Ndung'u and Waema, 2012; Ogutu, 2015). South Africa seems to suffer from policy inertia and has missed several opportunities to take the lead in this industry, especially in mobile telephony. It could be argued that the government has failed to harness this influential and powerful component of the economy since its inception.

One of the most irrational policy decisions the post-apartheid government made on mobile telephony, if viewed from a commercial standpoint, was the sale of its Vodacom shares. In response to a parliamentary question, the former minister of communications, Siphiwe Nyanda, stated that the sale of Vodacom was due to the shareholders' agreement that prevented Telkom from entering mobile telephony but allowed Vodacom to enter the fixed telephony market (Nyanda, 2009). The state initially had a considerable stake in Vodacom, a private company. In other words, the rationale was that to enable Telkom to compete in the mobile telephony market, they had to end the restrictions imposed by the Vodacom shareholders' agreement. However, Minister Nyanda's explanation is not convincing because if there was a genuine intention to make Telkom a significant player in the mobile telephony market, they would have needed substantial support to challenge the established players such as Vodacom and MTN. Knott-Craig argues that the government has made it difficult for additional MNOs, such as Cell C and 8ta (Telkom Mobile),[5] to compete with the duopoly of Vodacom and MTN (Knott-Craig, 2012: 33). He argues that the regulator could have lowered prices to foster competition.

Later in this book, I illustrate that mobile telephony was not only a prototype of post-apartheid business relations domestically but also a model

[5] Founded in 2010 and launched by Telkom as its mobile brand, it provided data and voice services on 2G and 3G networks, focusing on aggressive prepaid pricing to encourage usage. Later rebranded as Telkom Mobile in 2013, it is a subsidiary of Telkom SOC.

for African corporate expansion and value creation. South African firms tend to use mergers and acquisitions when expanding on the continent, allowing them to control 'capital invested' and influence policy direction (Luiz and Stephan, 2012: 625). In the case of mobile telephony, MNOs also built the supporting infrastructure and thus have control over the distribution channels. Moreover, in addition to complying with local regulations, South African companies export South African corporate governance culture as a condition for listing on the Johannesburg Stock Exchange (JSE) is following South African governance guidelines in their operations outside the country, such as the King Commission reports.[6]

[6] The King Commission reports are guidelines on corporate governance in South Africa, issued by the King Committee. The latest revision is King IV, published in 2016. Compliance is required for companies listed on the JSE. The King Code applies to all organizations and promotes ethical leadership. While voluntary, the reports provide valuable guidance for enhancing corporate governance practices.

3

South African corporate expansion into Africa

The role of capital in South Africa's position on the continent is central to its formation and statecraft. In this chapter, I examine the historical context of how big corporations, especially multinationals, have influenced South Africa's foreign and domestic policy apparatus. It outlines the historical dynamics and factors that shaped the nexus of mobile telephony and the State. However, this does not overlook the persistent ideological conflicts in state-business relations.

The dynamics of state-business relations in South Africa are not new; indeed, they are embedded into the foundations of contemporary South Africa. Colonies, as the country once was, were viewed solely as sources of resources and wealth, with no intention of self-governance. However, the narrative is incomplete without accounting for the role of private capital. Even during imperialism, the financial burden of expansion was beyond the means of colonizing states. This created a convergence of interests between corporates and the State, blurring the lines of authority. This is not to argue that contemporary multinationals are colonizing agents but rather to illustrate that the links persist and are not innovations of modernity.

I investigate the extent to which South African corporations have been influential actors in South Africa's foreign policy, as they have demonstrated in domestic policy processes. An important argument that cannot be ignored is that large corporates' human resource and capital maximization interests were an integral part of the reason apartheid was created and, ultimately, was weakened. This is indicative of the importance of big business to the State.

Charter companies: The original multinationals

Contemporary multinational companies wield considerable power, but this phenomenon is not novel. Sixteenth and seventeenth-century trading companies, such as the English East India (EEIC) Company and Dutch East India Company (DEIC), 'traded goods and services across national boundaries and had a geographical reach rivalling today's largest multinational firms' (Carlos and Nicholas,1988: 398). Arguably, charter companies were the original MNCs and gave expression to the home nation's largess. Let us look at what a charter is; it can be described as:

> a written instrument by which the State confers certain privileges on corporate bodies, either to protect them in the exercise of their lawful avocations at home, or else to encourage and sustain them in their more hazardous ventures abroad. (Cawston and Keane, 1896: 1)

They were companies created by an association of investors whose raison d'être was exploration, trade or colonization; they were often under the benefaction of one state. The relationship between charter companies and their home nation was tit-for-tat; the charter company was permitted to trade and explore, and in return, the company would *develop* and administer a territory on behalf of the home nation. This relationship imbued the companies with a degree of 'corporate sovereignty' (Kapferer, 2004) and paved the way for the entrenchment of companies in government, or even at times replacing state functions (Barkan, 2013; Davenport, 2013). Despite the decline of charter companies, contemporary MNCs still maintain a strong affiliation with their home state and act as their representatives (Kim and Milner, 2019).

The discovery of gold in Zimbabwe, Zambia and Malawi in 1886 bolstered the soon-to-be-created British South Africa Company (BSAC), which shored up Cecil John Rhodes's influence. The manner in which the Ndebele king, Lobengula, was swindled out of the mining rights in Matabele Land has been widely chronicled. It is understood that under the Rudd concession, Lonbengula gave up his tribe's rights in exchange for '£100 a month, 1,000 rifles, 100,000 rounds of ammunition, and an armed steamer on the Zambesi River' (Russell, 2009: 360). This became a model of how Rhodes would deal with native groups. He reportedly wrote to Queen Victoria bemoaning the tactics to secure the rights. However, Queen Victoria wrote back, basically telling him that the white people would continue to extract the gold, but his people and their property would remain secure (Ibid.: 34). After securing the rights to gold in Matabele Land, Rhodes made quick work of setting up the charter company – the BSAC, which Patrick Bond describes as being indicative of

> a structural switch from informal control of trade to trade with rule. British imperialists assumed that competition for control of Africa would continue beyond the 1885 Berlin conference which partitioned Africa, and that only BSAC-style 'imperialism on the cheap', as it was termed, would ensure geographical dominance over the interior of the continent in the face of hostile German, Portuguese, and Boer forces. (Bond, 2006: 94)

Thus, what we see from the BSAC example in southern Africa is a company being used to further the aims of the State and entrench cultural, political and economic leadership in the region. It may be argued that South Africa's significance, as an entity, in the region was gestated in the belly of its mines. In the growth of these extractive industries, we see a shift in the power of charter companies to the power of mining corporations. This was also around when 'virgin' territories were few and far between. Thus, capital garnered from the control of the Kimberley diamond mines made Cecil John Rhodes powerful, and valuable to the crown. He was the leading partner in De Beers that was the leading supplier of diamonds in the late nineteenth century, approximately 90 per cent of the world's supply (Russell, 2009: 32). Rhodes also developed the Consolidated Gold Fields of South Africa that had him involved in gold mining in the Transvaal.[1]

Antoinette Handley explains that economic and political power centred on race and ethnicity in developing the business class in South Africa between 1870 and 1989. Handley notes that the constant struggle between Afrikaner nationalism and perceived British imperialism was interjected by the struggle to control the extractive industries (Handley, 2008). The second Anglo-Boer War of 1899 to 1902[2] was primarily about control of these industries. This is

[1] Now called Gauteng under the democratic dispensation.

[2] The Anglo-Boer wars, fought between the British Empire and the 'Boer Republics' of South Africa, shaped the political and economic contours of the country. The first was (1880–1) centred on gaining independence for the country, and the second war, even though the British were the ultimate victors, began the end of British formal occupation of South Africa (Front-Line States, n.d.; Hopkins and Halstead, 2012).

demonstrative of the fact that state-business relations have always been fraught with tensions. However, it does not detract from the fact that there has been a degree of symbiosis or partnership.

In South Africa, the labour needs of the growing mining industry, as well as the burgeoning commercial agriculture, caused a situation whereby Blacks were left landless, and there was an increasing population of poor whites (Handley, Ibid.: 35). There was a strong leaning towards protectionism. However, the business community found itself divided on the issue. On one side of the fence, the import-export sector supported the free trade stance of the mining sector; on the other side was the agriculture sector which sided with protectionism (Ibid.: 39). Although the 1920s and 30s were marked by increased state involvement, Handley warns that the extent of state involvement should not be exaggerated (Ibid.). Evidently, the two state-owned enterprises, which drove much of South Africa's industrialization process, the Electricity Supply Commission (Escom[3]) and the Iron and Steel Corporation (Iscor), were largely independent and were not reliant on state funding. However, they were State supervised (Ibid.: 39–40). The International Development Company (IDC) was created in the same vein, with no access to public funding but with the intention 'to fund the establishment of private, not State, corporations in partnership with private businessmen' (Nancy Clark, Manufacturing Apartheid quoted in Handley, Ibid.: 40).

In the 1940s, as an expression of its power and extension of its reach, the State established several enterprises linked to the regime. These included Sasol, which manufactured and marketed fuel from coal; Safmarine, a shipping enterprise; Foskor, which dealt in phosphates; and Alusaf, which dealt with aluminium production (Ibid.: 41). Iscor and Escom began to operate as monopolies in their respective industries (Ibid.: 42).

What apartheid did domestically was ensure that the Black population was socially and politically forced into servicing white capital by providing cheap labour without any chance of contestation for the ownership of sources of capital. What is also clear is that there was a shift in the balance of power on two fronts: firstly, from British control to Afrikaner nationalists' control,

[3] It became known by the acronym Eskom in 1986 after the combination of the acronyms ESCOM and EVKOM – the Afrikaans name (Elektrisiteits Voorsienings Kommissie) of the power utility

and secondly, the State was using industry as an expression of its power and not the other way around: akin to the way imperialism made use of the charter companies. Yet charter companies, similar to future multinationals, also influenced the State. Archie Mafeje (1988) decried that the interests of the financiers of the State, or Big Capital, have always been paramount since the era of imperialism. He skilfully illustrates that the social stratification in the country was shaped to meet ever-evolving labour needs. However, he concedes that labour considerations are a mere piece of a larger socialization puzzle interspersed with the rise of Afrikaner nationalism and battles against British capital.

The South African regime established its own companies, such as Eskom, the IDC and Iscor. This could be seen as how the country ensured its influence over resources.

South Africa's foreign policy approach to Africa

Prior to the establishment of democracy, South Africa's foreign policy was driven by economic considerations and the quest for identity. Before the advent of true democracy in 1994, the literature reveals a struggle to determine whether South Africa was an extension of its colonial progenitors or an independent 'white' state within Africa. In response to UK prime minister Harry Macmillan's (in)famous 'winds of change' speech to the South African parliament, warning of the end of imperialism, Hendrik Verwoerd declared:

> We call ourselves European, but actually we represent the white men of Africa. They are the people not only in the Union but through major portions of Africa who brought civilisation here, who made the present developments of black nationalists possible. By bringing them education, by showing them this way of life, by bringing in industrial development, by bringing in the ideals with which western civilisation has developed itself. (Hendrik Verwoerd's response to Harold Macmillan's, n.d., 'Winds of Change' Speech)

This strong response brings two significant factors to the fore: by the 1960s, the ruling class in South Africa had a bipolar character as Europeans but also as the sovereigns of the rest of the continent. Adekeye Adebajo bluntly comments, 'patronising thinking was very much a feature of South African political thought from Cecil Rhodes to F. W. De Klerk' (Adebajo, 2017: 35). Indeed, the State's perception of the rest of Africa was informed by its identity politics and racial policies. Foreign policy and domestic policy are interwoven cords that make up the fabric of a state's overall approach. As such, there were intersections between the agenda of the State and that of the ruling political party. Moreover, big business's role in developing foreign and domestic policy cannot be ignored.

South Africa's neighbours were viewed as an extension of its territory – another source of cheap labour and raw materials. The words of imperialist and businessman Cecil John Rhodes best embody the general sentiment at the time:

> We must find new lands from which we can easily obtain raw materials and at the same time exploit the cheap slave labour that is available from the natives of the colonies. The colonies would also provide a dumping ground for the surplus goods produced in our factories. (Nau, 2014: 456)

As the State developed, it intended to expand its sphere of influence to include African countries outside southern Africa. Adebajo believes South Africa's contemporary view of the region as its backyard stems from the colonial era. Cecil John Rhodes was reported to have described the continent outside of South Africa's borders as part of its 'natural hinterland' (Adebajo, 2017: 35). One of South Africa's MNCs was the mining powerhouse De Beers founded by Rhodes. Patrick Bond contends that 'the consolidation of settler colonialism was feasible in large part thanks to the 1880–1890s entrepreneurship and geopolitical leadership of Rhodes's (Bond, 2004: 10). Bond further argues that:

> Rhodes's two main vehicles were the British army, which invented the concentration camp and, in the process, killed an estimated 14,000 black people and 25,000 Afrikaner women and children during the 1899–1902 Anglo-Boer South African War, and the British South Africa Company (BSAC), a for-profit firm which in 1890 began systematically imposing settler colonialism across the region. (Bond and Kapuya, 2006: 27)

After 1948, more specifically after the Second World War, South African corporations were more domestically focused (Siko, 2014: 135). In the post-World War era, South African firms started looking at opportunities outside

South Africa's borders with the assistance of groups such as the Federated Chambers of Industry (FCI) (Ibid.: 136). Rather than obstruct South African business, the 1950s fight for decolonization by African countries was a mere disturbance, partly owing to support from the South African government (Ibid.). South African companies traded vigorously in Rhodesia, colonial Mozambique, Angola and even francophone Africa (Ibid.: 136–7). At the height of the anti-apartheid movement between 1984 and 1988, African trade with South Africa grew from '6.5 to 10 per cent' (Ibid.: 137). To demonstrate the level of government involvement, Siko refers to the recollections of the former deputy foreign imnister (1986–9), Kobus Meiring, who made several clandestine trips across the continent to assist South African businesses in expanding (Ibid.). These efforts were successful as 'by 1990 South African firms "were quietly doing business with almost every country on the continent"' (Ibid.). Another example, from South Africa, of the involvement of the private sector in helping the State achieve its political aims is the case of Malawi and the Lilongwe Capital Project. Pretoria directly intervened in building Malawi's capital to make political inroads in Africa in the 1960s.

In 1966, the Malawi Development Corporation signed an agreement with the Johannesburg-based planning and design company, Imex (Pfister, 2005: 42). Imex only came into being a month before the signing of the agreement with the principal shareholders. They were Anglo-American, the IDC, Union Acceptances Ltd., and Anglo-American Private Merchant Bank (Ibid.). The agreement reportedly had a note stating that the agreement was 'expressly tied to the condition that maximum use was made of South African contractors' (Ibid.). Roberts Constructions Co.[4] was the leading company that won the contract (Ibid.). Pfister explained that the benefit of South African involvement was that the Department of Foreign Affairs was assisted in fulfilling its 'political ambitions', and for Roberts and Co., it landed a 'profitable and risk free order'(Ibid.: 43). Although, as conceded by Pfister, Pretoria only received a marginal benefit from this political and financial investment as Malawi was not a significant political player on the African continent, it did abstain from voting on the Lusaka Manifesto at the UN General Assembly[5] (Ibid.).

[4] This company merged with Murray & Stewart and became Murray & Roberts in 1967.

[5] The Lusaka Manifesto on southern Africa was voted for by the UN General Assembly, resolution 2505 (XXIV), on 16 April 1969. It spoke against white supremacy and oppression in southern Africa.

The relationship between the State and capital is interwoven but complex. When considering the sometimes hidden, embedded nature of the private sector in the State foreign policy machinery, it is essential to consider the context in which this occurred.

South African foreign policy towards Africa before the advent of democracy

The foreign policy ambitions of the pre- and post-democracy regimes have some notable similarities. The most striking are the State's aspirations for regional integration and leadership in the continent's emergence. Colonial interests ensured South Africa became a regional and continental power; the State enjoyed its inherited dominance. Early on, the State realized that its sustained economic and strategic relevance was linked to its external expansion. South Africa's immediate neighbours in southern Africa were subordinated to serve the interests of their colonial rulers; they were essentially 'Bantustans' (Nolutshungu, 1988: 161).

As Nolutshungu observes, 'Every government since 1910' recognized South Africa's economic supremacy depended on its expansion into Africa (Ibid.: 162). Organized businesses also played a significant role in identifying the economic opportunities on the continent (Adebajo, 2010: 104). Hendrik Verwoerd, the main architect of *grand apartheid*, claimed South Africa's ownership of the continent's potential (Adebajo, 2017: 35–6).

Before the democratic elections in 1994, South Africa's foreign policy had three dimensions due to its unique history. The first dimension was the State's relations with the West; the second dimension was the State's relations with its Black African inhabitants, whom it did not regard as citizens of the country; the third dimension was derived from the first and second dimensions – its relations with African neighbours.

Roland Henwood (1997) refers to the key features of South Africa's pre-1977 foreign policy as put forward by G. C. Olivier:

1. It was pro-Western and anti-communist. International trade was used to fight isolation and sanctions.

2. It strived for regional cooperation and 'South Africa as part of the African continent (with a permanent white population).'
3. It believed in avoiding intruding on the internal affairs of others to ensure a likewise response. Furthermore, using external justification to enhance South Africa's image overseas.

South Africa's shared values with the West, which stemmed from its similar histories of imperialism, resulted in many of its trading partners being Western nations (Adebajo, 2010: 104). South Africa had modelled itself around its identity as a stronghold of white power in the 'frontier' lands. However, the shift towards decolonization on the continent and the emergence of new actors prompted the government to react by becoming more oppressive (Nolutshungu, 1988: 162). The prospect of colonialism no longer being the norm in Africa understandably unsettled South Africa. South Africa's power, privilege and importance were based on the fact that it was under the patronage of imperialist powers (Ibid.). Decolonization represented a crisis for the State on both a domestic and international level (Ibid.: 163). Exploiting the hysteria of the Cold War era, apartheid leaders used the threat of the spread of communism in Africa to secure support from the American government (Ibid.).

After the Second World War, South Africa formed a relationship with European empires to bring 'development to the rest of Africa' (Ibid.: 105). In an apparent move towards regional integration, D. F. Malan in 1948 proposed an African Charter (Ibid.); his intentions, however, were not benevolent. The African Charter envisaged dual points of command: 'colonial powers and those powers with possessions on the continent' (Ngoma, 2005: 82). It blatantly sought to reinforce Euro-Christian hegemony in the region and prevent the spread of communism (Ibid.). Ngoma characterizes the African Charter and other initiatives, such as the Constellation of Southern African States (CONSAS), as South Africa's first attempt at building a regional security apparatus to perpetuate white supremacy (with South Africa at the helm).

South Africa's foreign policy underwent a metamorphosis from 1976 onwards, owing to several domestic and international pressures. The primary stages of this transmutation date to the Sharpeville massacre in 1960. This event was a critical turning point in South Africa's relations with the West and the rest of Africa. The massacre of sixty-nine people by the police was a result of the violent police response to the, by all accounts, peaceful protest, by Black marchers, against the use of a passbook. Even Western powers were uneasy about the situation in South Africa, which was largely ignored until then but was palpable. Less than a month prior to the Sharpeville massacre, The UK prime minister, Harry Macmillan, addressed the South African parliament and ominously warned them that:

> the wind of change is blowing through this continent, and whether we like it or not, this growth of national consciousness is a political fact. We must all accept it as a fact, and our national policies must take account of it. (Boddy-Evans, Harold Macmillan's 'Winds of Change' Speech online)

This speech has come to be seen as the recognition of the end of imperialism in Africa and 'the harbinger of an African nationalism sweeping irresistibly from the north' (Fleshman, 2010: 28). Being a South African ally was no longer fashionable.

The winds of change also blew over corporate codes of conduct. In the post-1976 era, the controversial Sullivan principles, devised by Reverend Leon Sullivan to respond to the growing demand for more Black participation in the South African economy, were the corporate code of conduct that promoted responsible corporate citizenship. These principles pertained to the conduct of US MNCs and the treatment of Black employees in the country (Lewis, 2001). One of the challenges of implementing the Sullivan principles was that it relied on self-reporting (Paul, 1986). Despite its limitations, the Sullivan principles contributed to the mounting pressure to end apartheid and laid the foundation for future corporate social responsibility philosophy (Stewart, 2011). In the immediate post-apartheid era, Black Economic Empowerment (BEE) emerged as a key economic-political strategy. The rationale of this strategy, as well as its later version of Broad-Based Black Economic Empowerment (BBBEE), was to provide preferential economic access for the previously disadvantaged.

The effect of African nationalism

As nationalism swept across Africa in the 1940s and 1950s, Pretoria became more oppressive in its tactics towards its population. G. C. Olivier argued that the changing external dynamics led to widespread support, from both English and Afrikaans-speaking white South Africans, for Hendrik Verwoerd; his solution to the problem was to re-engineer the State (Olivier, 1982: 274).

This period saw legislation against racism in the country. The apartheid administration built on older racist legislation, such as the 1913 Native Land Act, and enacted several pieces of legislation to formalize and intensify racial discrimination that had begun in the seventeenth century: legislation such as the *Bantu Homelands Constitution Act 21 of 1971* and the *Bantu Affairs Administration Act 45 of 1971*. The South African State after 1948 was repressive, as it had devised a system that relegated Black Africans to be an essential part of the functioning of the State but not recognized as members of the State. This situation was somewhat akin to the Roman citizenship laws, where different classes of people had varying categorizations within the State; in the case of South Africa, Black Africans were never meant to be able to attain citizenship. Black Africans were regarded as foreigners within South Africa. The government achieved this state of affairs by creating the homeland system known as Bantustans.

Ten states within the State were designed to create 'elf-governing' ethnic holdings. The spatial development of these holdings meant there was always a dependence on the larger South Africa. Schrire noted that the homelands depended on their ties to South Africa for employment and product access. Conversely, South Africa's core economy depended on the homelands to provide labour (Schrire, 1982: 136).

The National Party government perpetuated the narrative that these homelands were, in fact, the traditional tribal lands of Black Africans. Under Verwoerd, separate development, or apartheid, was conceptualized. D. F. Malan promulgated the Bantu Authorities Act in 1951 to establish various ethnic reserves for the country's Black population. In 1959, the Promotion of Bantu Self-government Act was passed, and the actual plan for apartheid was detailed. Ultimately, the act created the following African reserves: Transkei (1963), Ciskei (1972), Le bowa (1972), Venda (1972), Bophuthatswana (1972), KwaZulu (1973), Gazankulu (1973), Qwaqwa (1974), KwaNdebele (1977) and KaNgwane (1984). Four of these states were later proclaimed independent states. These were the so-called TBVC States – Transkei (1976), Bophuthatswana (1977), Venda (1979) and Ciskei (1981).

In 1970 the citizenship rights of black South Africans were removed through the Black Homelands Citizenship Act. Robert Schrire describes the primary rationale on the part of the State as follows:

> Each individual black, irrespective of culture, birthplace or residence remains an immutable component of one these ethnic nations. The aim of the policy is, in its simplest terms to disentangle South Africa's multiracial society which past economic and social forces have created. (Schrire, 1982: 114)

Schrire contends, however, that these homelands did not match the persistent narrative of being traditional homelands but, more accurately, were 'the areas of black occupation that remained after the military and political defeats of various African communities and kingdoms' (Ibid.: 115). He further contends that the 1913 Land Act and the 1936 Natives Representation Act were designed to segregate white and Black people to serve 'the needs of white farmers and industrialists for black labour' (Ibid.).

Naturally, these homelands persisted because of the emergence of a Black ruling class that benefited from maintaining the status quo. Diplomatically, the only country to ever fully recognize these homelands was South Africa itself. Yet, it is not plausible that the government intended these states to be independent. The homelands becoming sovereign nation-states would have reduced South Africa's power. The South African government's military intervention in Bophuthatswana's military coup in the 1980s was indicative of Pretoria's willingness to use state resources to protect not only its interests but also those of capital; Pretoria restored Lucas Mangope a mere fifteen hours after he was ousted on 11 February 1988 (Battersby, 1988). Not only did South Africa's security forces descend on Bophuthatswana, but President P. W. Botha, defence minister Magnus A Malan, law and order minister Adriaan J. Vlok and foreign minister Roelof F Botha flew to the homeland's capital that same day (Ibid.).

It has been widely reported that there had been similar coups in the Transkei six months prior, but the response was not comparable. This was probably because Bophuthatswana was home to the Sun City Entertainment complex and significant platinum reserves. Pretoria again intervened in 1990 following civil unrest in the homeland on 8 March 1988 (Wren, 1990).

South Africa and the rest of Africa

The State could be described as malevolent towards countries that were not so far from its borders. The establishment of Southern African Customs Union (SACU) may be considered evidence of its malign intent. Instead of being a tool to integrate regional economies, it was an institutional apparatus that sought to deal with the economic integration of separate political dependencies (McCarthy, 2008: 14). This meant that the countries within the southern African region became an accessible market for South Africa's goods as well as easy access to resources – human as well as commodities. 'In the most hegemonic of CUs,[6] SACU, South Africa simply decided trade policy and compensated the smaller countries for the costs it imposed on them' (Schiff and Winters, quoted in McCarthy, 2008: 16).

The growing internal and external opposition to its policy of apartheid marked the relationship between South Africa and the rest of the continent. The country became increasingly isolated but maintained a stranglehold on the region through SACU. The frontline states, southern African states opposing apartheid, were formed in 1970 to harmonize their responses. These states were Angola, Botswana, Lesotho, Malawi, Mozambique, Swaziland, Tanzania, Zambia and Zimbabwe. They found it challenging to pursue their agenda as they were all economically dependent on South Africa. This led to the formation of the South African Development Coordination Conference (SADCC) on 1 April 1980 to counterbalance this economic strength.

> The formation of SADCC added to the isolation of South Africa from the international community when more European countries, including the United States increased their support for the SADCC. (Front-Line States, n.d.)

Even though Pretoria found itself increasingly isolated as the opposition to apartheid grew, there were apparent attempts to diplomatically engage with the rest of the continent: for instance, J. G. Strydom was one of the first heads of state to congratulate Sudan on gaining its independence, and in 1957 there was a diplomatic representative at Ghana's independence celebration (Adebajo, 2010: 104).

[6] Customs Unions.

According to Adebajo, racism in the country and the African governments' anti-apartheid stance were the main reasons these burgeoning relationships with newly independent African nations could not be deepened and leveraged (Ibid.). Foreign policies in the region were often 'tools of managing a hostile regional environment' (Khadiagala, 2001: 132). In other words, African countries' diplomatic actions had an element of hegemonic response strategies embedded in them. With the development of SADCC and the creation of various coalitions at the time, they could lessen South Africa's power to a degree and ultimately bring apartheid to an end.

South Africa's attempts to carve out a new role

The integration of the 'new' South Africa into the sub-region as well as ensuring its survival was a subject of great debate by other African countries. Understandably, it elicited concern from statesmen on the continent. For instance, at the African Leadership Forum held in Namibia in 1991, wearing the cap of chairman, former Nigerian president, Olusegun Obasanjo, noted that:

> in economic and social terms, South Africa is weak and unwell. The situation is likely to get worse before it can get better as a result of the transition and the uncertainty of the interim situation. With the demise of apartheid and the correction of gross misstructure and the underperformance of the economy, the social malaise brought about by the policy of apartheid will remain for some time. Deliberate efforts will have to be made to deal with the social problems of the 'new' South Africa. It is now being recognized by all concerned in South Africa that the problem is not that of changing the captain of the boat but changing the engine and reorganising the crew. That reorganization may inevitably lead to a change of the captain but, all in all, fears will have to be allayed and expectations will have to be moderated. (Obasanjo, Chairman's Summary: Conference on the Challenge of Post-Apartheid South Africa to Southern Africa in particular and Africa in General, 8–10 September 1991)

Lesotho's King Moshoeshoe II revealed the fear of many countries in the front-line states about the potential danger that southern Africa will ' continue to be subordinated to South Africa' (Moshoshoe, 8–10 September 1991).

Sanusha Naidu describes a few of the factors at play at the time that South Africa undertook its political transition: first, the international political economy and global architecture were changing; second, as a result of the change, the relationship between 'countries of the Third World and the Cold War architects' also underwent a metamorphosis; and third, these changes impacted 'the political architecture on a domestic level' (Naidu, Skype Interview, 2016). In other words, because of the political shifts that had taken place at the global level, the political environment had become very different. This meant, in practical terms, that the near-simultaneous ending of the Cold War as well as the end of apartheid opened the way for South African corporations' expansion on the African continent. Ultimately, South Africa's reintegration into the international system resulted in an increasing economic shadow over the continent and the sub-region that fuelled fears of the new South Africa's dominance.

South Africa's attempt to create a role on the continent and on the global stage has always been a matter of much contention. Among technocrats and academics alike, there was widespread recognition that the success of the new South Africa was vital to the continent.

Consequently, a debate raged in academic circles on what direction South Africa's foreign policy should follow. In the post-apartheid era, relations with the rest of the continent have been undeniably shaky.

Conclusion

Capital, particularly MNCs, have long been part of the fabric of the South African state – during and after apartheid. In the immediate aftermath, there were several expectations of the newly installed ANC government. The new democracy in South Africa had high hopes to fulfil. Domestically and internationally there was an expectation that Nelson Mandela and his ANC government would be a model for Africa – a champion of freedom and human rights, a healer of past wounds and an economic powerhouse.

These dynamics affected state-business relations, hampering them with ideological tensions. The irony is that the old and the new South Africa had a messianic complex. However, it is undeniable that the end of apartheid opened new opportunities. For example, the country rejoined the United Nations and the Commonwealth. It also became part of the organizations that opposed the apartheid regime – Organisation of African Unity (OAU) and the Non-Alignment Movement (NAM). At the same time, the new South Africa inherited the old South Africa's debt and economic stagnation. Mills (1997) argues that Pretoria lacked a clear strategic direction in its post-1994 foreign policy and based it on a broad list of ideals. This led to a dilemma where South Africa wanted to help its neighbours but avoid being too dominant; in other words, Pretoria wanted to lead from behind.

Ultimately, as Southall posits (1994), the simultaneous collapse of communism and the end of apartheid presented South Africa with a 'double whammy' of challenges, the most significant of which was balancing internal and external political forces while developing a new foreign policy. Southall's primary focus was on achieving and maintaining democracy. His analysis is based on three 'camps': the 'nationalists' camp, the 'economistic' camp and the 'open society' camp.

The first camp believed South Africa would adhere to democratic ideals due to its 'international standing' (Southall, 1994: 124). The second camp assumed a future democracy and did not consider the 'political nature of "post-apartheid" South Africa' (Ibid.: 125). The last camp, primarily composed of South African businesses, believed in the link between a free market economy and democracy (Ibid.: 127).

It did not help that South Africa's foreign policy orientation over this period was unclear as, over the years, South Africa's foreign policy formulation has experienced several trials and errors. However, it has evolved from the uncertainty that marked the foreign policy between 1994 and 1999. During this time, the direction of foreign policy was derived from speeches, discussion documents and so on. It is worth noting that the principles now clearly outlined in the Department of International Relations and Cooperation (DIRCO) report are not significantly different from those in the 1994 foreign policy discussion document presented by the then minister of foreign affairs, Alfred Nzo.

Thanks to the ultimate neoliberal thrust, South African corporations have expanded their influence across South Africa's borders across its border. The next chapter will further explore South Africa's corporate expansion and what it means for South Africa's position on the continent.

4

Corporate expansion in the democratic era

The telecommunications sector was the first economic sector to undergo reform in the new democratic South Africa, and it became a 'model' for other sectors (Horwitz, 2001a: 18–19). Besides providing services for previously underserved Black communities, the telecommunications sector also saw one of the first offerings of BEE shares.[1] The telecommunications sector offered a prototype for South African corporate expansion in the post-apartheid era, even though it was not the first sector to expand. This model is most evident in mobile telephony.

Understandably, mobile telephony was an accessible vehicle for a new form of corporate engagement: it offers to quickly close the long-standing communication infrastructure gap; it also offers more accessible avenues for poorer African countries to participate in the internet age. Telephony and internet access allow Africans to innovate and participate in the global economy. Examples of African-born innovation are products such as the mobile money transfer service M-Pesa (Davidson, 2012: 5) and the development of innovation incubators such as the iHub in Nairobi, Kenya (Puhl, 2013); Technopark in Stellenbosch, South Africa (Mahlaka, 2015); and the Tshimologong Digital Precinct in Braamfontein, South Africa (Tshimologong, 2016). In 2016 alone, there were 314 active Tech Hubs in Africa, with MNOs partnering with about 13 per cent of them (GSMA, 2016). For instance, MTN has supported the launch of the Activspace hub in Cameroon (Dahir, 2016) as well as Jumia (formerly Africa Internet Group) in Nigeria (Akwaja, 2013).

[1] However, the partial privatization of the State telecoms company, Telkom, to create the Khulisa scheme only resulted in less than 1 per cent ownership of Telkom being transferred to select Black people (Rumney, 2005: 407).

The potential offered by changing technologies parallels South Africa's changing ideological orientation in the democratic dispensation. They are steering towards closer relations with other developing countries and fostering an environment of self-reliance.

In this chapter, I evaluate South African corporate expansion after 1994. As democracy emerged, the state's relationship with corporate powers changed as it became more concerned with being seen as an outcast on the rest of the continent. At the same time, South Africa's status as a fully inclusive democracy opened the door for outward investment. As well as giving a brief overview of inter-African trade in communications-related technology, I examine South Africa's foreign policy regarding its corporate interests. I only focus on one part of the mobile phone production network: MNOs. This is because South Africa's entry point into mobile telephony was through MNOs, not Coltan mining or equipment manufacturing. Analysing the rise of South African MNOs shows how they achieved the status of MNCs and became politically influential.

Mobile telephony MNCs and the question of South Africa's ambitions

The main reasons MNOs expand in Africa are the same as in other sectors: they want to grow, make money and stay in business for a long time. These companies are not primarily interested in politics but sometimes get involved because of their expansion. In this chapter, I argue that MNOs help South Africa to show its soft power and leadership in the region and the continent. They do this in two ways: by spreading South Africa's influence and getting other countries to accept its role.

South Africa's corporate expansion helps to foster stronger ties, but the country does not have effective ways of coordinating this expansion. As economic relations deepen, political partnerships become more critical, and cultural integration happens naturally.

To better understand the operating context for local and foreign firms that use South Africa as a platform for expanding into other African countries, it is important to briefly discuss foreign MNCs with base operations in South

Africa. Peter Drucker (1960) characterizes the MNC as 'a tool and organ of society'. This applies to both South African MNCs operating abroad and foreign MNCs operating within South Africa (including those using the country as a gateway to the rest of the continent).

MNCs have to comply with the laws and policies of the host country. In South Africa, a foreign company can operate in two ways: first, it can register its company with South Africa being a branch of the international parent body; a 'Memorandum of Incorporation' has to be registered in South Africa. Second, the MNC can open a local subsidiary; this means that the company opens a private South African company in which shares are owned by the MNC.

Some examples of international telephony companies that have invested in South Africa are French Telecommunications company Alcatel-Lucent, through local partners Altech group and the Black empowerment group Rethabile; American Cisco Systems, which has a history in South Africa dating back to 1989, partnered with Dimension Data; and British Vodafone, which controls a significant stake in Vodacom. Foreign registered companies or local subsidiaries must operate according to the BEE charter, although BEE ownership targets remain contentious (Bowman Gilfillan, 2004). The same applies to BBBEE, albeit with some flexibility in the ICT sector (Sánchez, 2008: 111). The ICT sector's flexibility was because the equity partnership requirements were not as stringent as in other sectors. In 2016, an amended BBBEE ICT sector code was launched. The new code gives more weight on the BEE scorecard to various categories of Black people such as women and the disabled. The areas given priority in the code are ownership, supplier and skills development (Parliament of the Republic of South Africa, 2016: 13).

Despite the administrative burdens of operating within South Africa, it remains quite attractive to foreign MNCs and provides a context for discussing South African MNCs that operate beyond its borders. For all MNCs, especially in the mobile telephony sector, there are advantages to being among the first movers.

Diana Sánchez's study examines how South Africa's transformation agenda has shaped the operations of MNCs within and outside the country. The case study used is that of Ericsson, a Swedish telecommunications MNC that uses South Africa as a regional hub for its expansion in the rest of the continent (Sánchez, 2008: 105–23). The rationale for choosing Ericsson is its long history in South Africa in both the pre- and post-apartheid era; it is a major infrastructure provider for MTN and has a growing market share in Africa. Sánchez argues, 'The story of Ericsson in post-Apartheid South Africa is illustrative of how MNCs are responding to South Africa's Broad-Based Black Economic Empowerment (BBBEE) strategy' (Ibid: 105). The author reveals that MNCs circumvented the BBBEE requirements by having two headquarters in South Africa: one for the local market and one for the regional market. The local headquarters would comply with the local laws and transformation processes (Ibid: 106).

However, the question remains: how much benefit does the rest of the continent derive from having the headquarters based in South Africa?

Who serves whom?

As discussed previously, state-business relations entail that the state creates the conditions for businesses to thrive both at home and abroad through a stable economy, attractive trade agreements and favourable state-state relations. South Africa's history is intertwined with its relation to capital (see Chapter 3); however, this is not peculiar to South Africa. The problem arises when a domestic MNC's conduct abroad tarnishes the country's image.

In a study on South Africa's perception abroad, Neuma Grobbelaar found that South African companies are seen as aloof, potentially exploitative and having an unfair competitive edge (Grobbelaar, 2006: 68–9). However, these perceptions are not based on the actions of individuals or small firms and the possible adverse effects of increased trade. Brand SA field research in Nigeria and Kenya found that South African companies are not only preferred employers but also enhance the culture of corporate governance (de Kock, 2014: 4). However, they also found that South African companies are aggressive and sometimes disregard the advice of local advisors (Ibid: 3).

Regarding corporate governance, South African legislation, especially the South African Companies Act 71 2008 section 72 and regulation 43, requires companies of a specific size listed in South Africa to have social and ethics committees that oversee corporate responsibility and corporate governance. The King Report III lists corporate responsibility as one of the code of conduct principles. Companies listed on the JSE have to apply the code. But they are not obliged to strictly follow the code because of the 'apply or explain' principle which means that a company must apply the code of conduct or explain in their annual reporting why they deviated from the code. The King IV Report (King Commission, 2016), which will replace the previous King codes, has abolished this principle.

Beyond what was expected of them in the mid-1990s, MTN's case shows that the company was very conscious of the impression the company created by the poor behaviour of expatriate staff. Thomas Lutete Maketa demonstrates the extent to which MTN was concerned about the impression it created. He provides an example of racial incidents involving white South African staff in its Nigeria operation. MTN dealt swiftly and decisively with these incidents to avoid headlines such as: 'South Africans bringing apartheid to Nigeria'(Maketa, 2009).

There is a complex interplay of interests between the state, its domestic MNCs and the host country. A US example by Jade Miller looks at the different interests by examining the February 2006 congressional hearings into four US MNCs' actions in China. Google, Microsoft, Yahoo and Cisco were accused of aiding the Chinese government. Miller notes that the United States has a long history of supporting industry, especially for expansion beyond US shores (Miller, 2009: 287). The United States can assist its domestic MNCs and purely local private industry through state-state trade agreements (Ibid: 288). However, it struggles to rein in MNCs for behaviour against its national value system. Miller argues that in these situations, the state has to resort to 'soft power' approaches to elicit cooperation. The hearing into the big tech MNCs highlighted the state's willingness to confront its domestic corporations but 'at the same time not serve to disrupt the success of American MNCs in the world order' (Ibid: 300).

The case of mobile telephony in South Africa is particularly complex. There have been grievances from policy practitioners and academics about the lack of support for the business sector as a whole. Despite the calls for a uniform regulatory body for South African businesses operating abroad, no such body has been established. The experience of MNOs is more nuanced, perhaps due to their unique role in South Africa's history (as discussed in Chapter 3). The ruling ANC, at its fifty-second party conference, also advocated for a code of conduct for business (ANC, 2007).

South Africa has the largest telephony market on the continent, with very high penetration rates. As a result, the companies emerging from this market are advanced and equipped with the capacity to cope with the additional challenges often encountered on the continent. On the continent, South Africa dominated the market for telephony and ICT. Besides infrastructure and human capital, the policy environment was part of the attraction. However, success in implementation sees South Africa falling behind. This situation can be attributed to two main factors: lack of leadership and scarcity of implemented policy. Between 1999 and 2014, six changes to the leadership of the Department of Communication resulted in missed deadlines and insufficient momentum for great ideas.

Moreover, South African MNOs operate in an environment that lacks a coherent code of conduct. This absence of a clear policy and regulatory framework increases the risk of corruption and other unethical business practices. Ewan Sutherland found that the telephony market and its associated policies increase the risk of corruption (Sutherland, 2012). For example, former Sierra Leone president Charles Taylor was accused of using Lonestar, the former monopoly mobile operator in the country, as a source of personal income by granting them an exclusive license (Carvajal, 2010). Lonestar was later acquired by MTN.

Another risk for all MNCs is being viewed as proxies by their host nations and can quickly become targets. For instance, following the 2014 outbreak of xenophobic violence in South Africa, South African MNCs faced increased pressure and scrutiny. However, some companies have adopted a localization strategy to create a sense of ownership of the local subsidiary of South African companies. Petrus de Kock notes that Stanbic ITBC (Standard Bank International Trade Business Centre) has been particularly successful in these efforts (de Kock, 2015).

Conclusion

The mobile and information economies are very linked in sub-Saharan Africa. Mobile phone MNCs have an advantage in the region because of the lack of infrastructure and other reasons. The GSMA said that by the middle of 2014, there were about 600 million SIM card connections in the region, which means a penetration rate of 68 per cent (GSMA Intelligence, 2014: 8). This has a big impact on the region's economy, as the whole mobile phone ecosystem adds to GDP directly and indirectly. The rise of MNCs from emerging markets has strengthened relations between emerging market economies. The stronger relations have happened simultaneously as, and maybe helped to cause, stronger political relations. The result is that emerging markets have not been hurt as much by the 2008 financial crisis that affected the global economy (Africa Progress Panel, 2012). Some of the best emerging market MNCs are from South Africa. The main question in this chapter was if South African MNCs in the mobile phone sector have strengthened South Africa's leadership. Even though MNCs seem to have a lot of influence, there is no clear proof that MNCs directly change how the South African government acts. How South African companies act abroad matters for how people see the South African government and its public diplomacy efforts. Also, on a local level, the South African government needs its MNCs to do well outside the country because their profits add a lot to the South African GDP.

5

Vodacom

Vodacom's case illustrates the complex and contradictory dynamics of telecommunications in Africa, where state intervention, foreign investment, market regulation and technological innovation interact in shaping the development and governance of the sector. Its story reflects the shifting dynamics in the relationship between the state, the market and society in the context of transformation. The initial duopoly of MTN and Vodacom in South Africa did not emerge in isolation. As detailed in previous chapters, in the lead-up to 1993, when the first cellular telecoms licences were awarded, they were shaped by several shifts in the country's political economy. Moreover, the national government's macroeconomic strategy played a large part in the future dominance of South Africa's founding mobile network operators. The march towards market liberalization as well as the transformation of the regulatory environment were reflections of South Africa's dynamic political environment and changes in the global telecommunications regime.

As discussed in Chapters 1 and 3, South Africa's changing political environment was interlinked with attempts to change the economy – and South Africa's positioning in the world.

In this chapter, I examine and trace the history of Vodacom in South Africa, still a leading mobile telecommunications company in the country. Vodacom originated from a partnership between the state-owned fixed-line operator Telkom and the UK-based multinational Vodafone, emblematic of a particular form of the government's BEE strategy. However, over time, Vodafone increased its stake in Vodacom, raising questions about its alignment with South Africa's foreign economic policy and domestic objectives. I analyse how Vodacom's ownership structure, market position, regulatory environment and social impact have evolved and shaped its role in the country's mobile telephony sector. The story of Vodacom illustrates three key lessons: the limits of transformation, the shifting balance between states and markets and how companies can become co-opted into the internationalization agenda of the state

The history of Vodacom

Vodacom emerged in the early 1990s as a joint venture between the South African state-owned Telkom and the UK-based MNO Vodafone, becoming the country's first and largest mobile operator. The choice of Vodafone as an international partner was strategic for a number of reasons; first, Vodafone was behind the introduction of mobile telephony in Britain; second, it launched GSM in 1991, in comparison to other competing MNOs at the time; and third, its expansion strategy hinges on always marshalling its first-mover status with regards to the latest technologies. For transitional South Africa, the decision to partner with a company with network experience from similar corporate culture made the politically fraught decision a little easier.[1]

The fact that GSM was the standard that prevailed after the back-and-forth horse trading between the regulator and the transitional government additionally made Vodafone a strategic partner. Lastly, Vodafone has a history of innovative product offerings, it was the first to introduce data, fax and SMS to its customers in 1994 (Anwar, 2003).

As a result, Vodacom also had the benefit of first-mover status, which helped it cement its place as the most successful MNO regarding subscriber numbers. Ultimately, the most remarkable benefit / or lesson that Vodafone's partnership passed onto Vodacom was *how to build an empire* (Curwen and Whalley, 2004).

Vodafones's partnership with Telkom to create Vodacom could have easily come straight out of Vodafone's playbook in creating a corporate empire. Vodafone now has a presence in almost every country. Its primary internationalization vehicle is acquiring stakes or licences in other country's

1 Like South Africa, Britain, had a fixed line monopoly. The introduction of mobile telephony was a political hotcake, with unions, and regulators, all having a say.

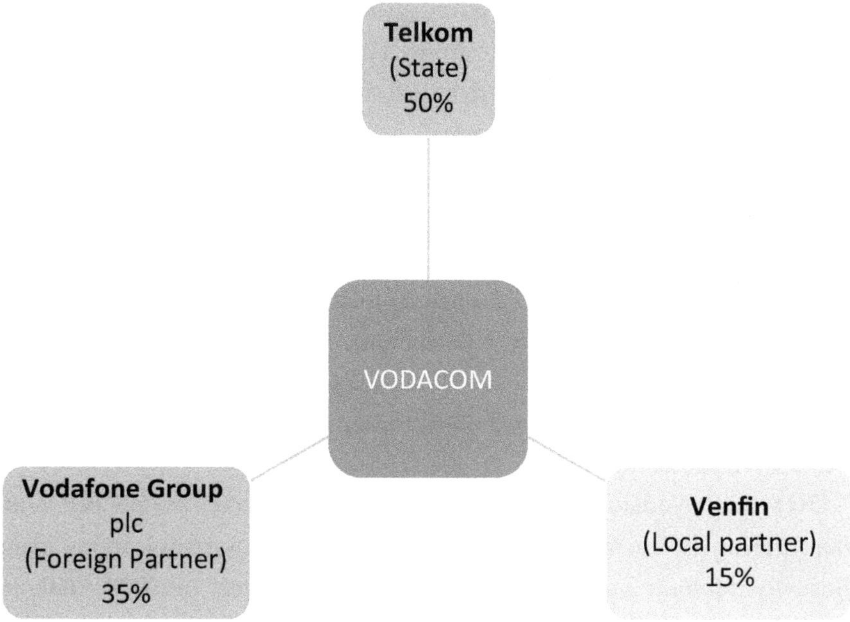

Figure 2 The Vodacom ownership structure 1994.

Source: Author's interpretation.

mobile networks. In 1993, it formed the Vodafone Group to obtain overseas partnerships starting with Germany, South Africa, Australia, Fiji and Greece (Anwar, 2003). The same modus operandi drives Vodacom's contemporary internationalization process.

Vodacom was incorporated in 1993 as a joint venture between Telkom (50 per cent), Vodafone (35 per cent) and Venfin[2] (15 per cent).

Vodacom (see Figure 2) initially did not have an empowerment partner. In the aftermath of the CODESA breakdown, and the NP government going ahead with its mobile telephony plans without consultation, a deal was struck with regards to the license holder composition of MTN and Vodacom. For MTN, it was agreed that 30 per cent of its equity will be ring fenced for Black shareholders, while Vodacom would give 5 percent of its shareholding and provide other opportunities 'downstream' (Cawe, 2020: 250). This was one

[2] Then a venture capital subsidiary of Rembrandt Group, the conglomerate was founded by Afrikaner businessman Anton Rupert in the 1940s.

of the first BEE deals of the new government. In 1996, Vodafone and Venfin sold a 5 per cent stake to a BEE group, Hosken Consolidated Investments (HCI), founded by the South African Clothing and Textile Workers Union (SACTWU) secretary general, Johnny Copeland.

In 1997, HCI was acquired by the Mineworkers Investment Company (MIC) and SACTWU Investment Group, ultimately the forerunner of 'broad-based black economic empowerment' (Rose, 2008).

In 2002, HCI made a killing when it sold the 5 per cent stake back to Vodafone and VenFin for over a R1 billion; Venfin got 1.5 per cent of the stake, and Vodafone got 3.5 per cent (Fin24, 2002). In 2006, Vodafone bought VenFin's stake and grew its shareholding to 50 per cent (Wray, 2007; 'Vodafone in South Africa', 2008).

Over time, Vodafone increased its stake in Vodacom to 65 per cent, transforming it into a subsidiary of the global telecommunications giant (Business Monitor International, 2009: 45). However, in 2009, Telkom concluded the process of divesting its interest in Vodacom to Vodafone (Jurgens, 2011: 24). Vodacom, via Telkom, provided the State with a significant amount of revenue. Although, as reported by Africa & Middle East Telecom-Week online, there were questions about untoward deals with ICASA and possible price fixing (Africa & Middle East Telecom-Week, n.d). This divestment also deprived the state of a valuable source of influence over the sector, as the government sold its remaining shares in Vodacom to the public investment corporation (PIC) at a discounted price. The Treasury, led by the minister of finance, sold a stake in a state-owned asset to the PIC, also state-owned and chaired by the deputy finance minister. Despite the 10 per cent discount received by the PIC, this transaction potentially jeopardizes public service pensions, as the PIC is responsible for managing these funds. The purchase of a stake in a struggling foreign company subsidiary may compromise the PIC's ability to fulfil its obligations.

Interestingly, Vodacom's market entry in territories North of the equator in Africa was restricted by a shareholders' agreement between Vodacom and Vodafone (Business Monitor International, 2009: 45; Balancing Act Issue 329). Vodacom Group is made up of Vodacom South Africa (SA) as well as its other international holdings. In 2008, Vodacom Group divested part of its 100 per cent share in Vodacom SA in a BEE deal with subsidiaries of Royal Bafokeng Holdings (Proprietary) Limited (Royal Bafokeng) and Thebe Investment Corporation (Proprietary) Limited (Thebe) including share schemes aimed at increasing the amount of Black ownership (Business Monitor International, 2009).

The selling of the government's remaining shareholding in Vodacom Group, coupled with Vodafone no longer having restrictions on Vodacom's African expansion, means that the group's presence on the continent is set to grow. However, the government's unbundling of more of its stake in Vodacom is puzzling on several counts. Firstly, though it is branded as a South African company, Vodacom is really a subsidiary of a British company. The unbundling of direct government stakes in the enterprise, perceptually, gives the impression that any potential reins the government had as a significant shareholder in the business have diminished.

However, Vodacom's integration into Vodafone's network enabled it to access new markets and technologies while also constraining its strategic autonomy and competitiveness. Initially, Vodacom was barred from entering territories where Vodafone had existing operations, limiting its expansion opportunities in Africa. Although this restriction was lifted in 2009, Vodacom faces stiff competition from other regional and global players such as MTN and Orange. Furthermore, Vodacom's acquisition of South Africa's second fixed-line operator Neotel raised concerns about its potential monopoly over spectrum and infrastructure and its impact on consumer choice and affordability. In 2018, Vodacom was made a stand-alone business to drive its African expansion. This shift in ownership has significant consequences and implications for the company and the sector.

Vodacom and the state

An enduring thread with state-business relations in post-apartheid South Africa has been concerns around how to increase Black ownership of the economy whilst ensuring increased service coverage to the most marginalized in society. One of the fast mechanisms to change the economy, and to increase Black ownership, has been through BEE partnerships. Yet, over time, the quality of many of these partnerships has been called into question for a number of reasons, including the inability of many Black investors to sustain their shareholding and opaque BEE shareholding schemes that did not give Black shareholders actual ownership (Cawe, 2020; Levy et al., 2021).

The threads of inclusion and redress appear interwoven into the objectives of the state and the internationalization mission of many large MNCs, particularly MNOs. South Africa, mainly via its corporate sector, needed to break with the negative vision from its past; corporations need to expand to remain viable. In the case of Vodacom, its continued operations in the country – and state support in its expansion efforts – cannot be wholly divorced from BEE concerns. Yet, many BEE deals, including those in telephony, have been negatively affected by changes in the political climate.

For foreign companies entering the country, particularly during the early stages of political transition, the ability to leverage the post-apartheid state's outreach into the global community was crucial. This included capitalizing on the goodwill the ANC government had garnered through its successful transition from an authoritarian, segregationist regime to a fully inclusive democracy.

For Vodafone, the market conditions were ideal for expansion, and the 'new' South Africa presented a perfect launch pad for their services throughout the continent. In 1994, Vodafone acquired a cellular license in Uganda when Vodacom South Africa's services went live. Their foray into South Africa substantially exceeded expectations, capturing 65 per cent of the market by the end of 1994 (Vodafone Group Plc, 1995: 14). By 2003, despite the entrance of another competitor, Cell-C, Vodafone still maintained a 58 per cent market share. This was due in no small part to their offering of prepaid services, which accounted for 85 per cent of their customer base (Vodafone Group Plc, 2003: 14).

Empowerment to improve state-business relations

A challenge Vodacom has had, which is not unique, is partnering with a suitable empowerment vehicle so as not to run afoul of the state whilst still meeting its obligations to its shareholders regarding growth and profitability. However, as discussed in Chapters 1 and 4, the racial makeup of capital in South Africa meant that the ANC government had little choice but to intervene to ensure that Black South Africans were no longer edged out of control of the formal economy. The response was also informed by witnessing the success and failures of economic decolonization in other African counties. As Roger Southall (2006: 3) points out, the ANC had the benefit of witnessing the flaying attempts of other decolonizing regimes to take over their economies via nationalization, with often disastrous consequences, and they also had a first-hand seat to the workings of Afrikaner nationalism that seeped into the economic agenda of the NP government. Understandably, the nature of post-apartheid capitalism meant that an almost unholy alliance had to be struck between business and the ANC elite. Southall best explains:

> The complement to 'big business' securing the turn in ANC economic policy was the need to embrace the incoming ANC elite into the direct sphere of capital. The approach was broadly two-fold: increasing the recruitment of blacks into corporate employment as managers and board members; and facilitating the rapid expansion of ANC-related black-owned business. (Ibid.: 5)

Mobile telephony companies, including Vodacom, have had to navigate the political exposure of empowerment schemes while fulfilling their transformation obligations in a state where most of the population are still excluded from the formal economy. Despite its growth, Vodacom still struggled with getting an empowerment partner or vehicle, as expected under the BEE requirements of the country. Part of the problem with the empowerment requirement, though necessary for social and economic redress, is that it makes room for political gerrymandering and other reputational risks. Such was the case with the Elephant Consortium's bid to become Vodacom's empowerment partner in 2007 prior to its listing on the JSE.

To understand the nature of the risk, one would have to turn back to 2004, when the Elephant Consortium bought Telkom's empowerment shares – subsidized by the PIC. Exposure of the shareholding caused an uproar as many of the shareholders were influential members of the ANC and its breakaway party, Congress of the People (COPE). Beneficiaries of the deal include Smuts Ngonyama,[3] Gloria Serobe,[4] Wendy

[3] Former South African ambassador and former head of communications for the ANC. At the time of the Elephant Constortium scandal, he was one of the founding members of COPE.

[4] Cofounder and CEO of Women Investment Portfolio Holdings (WIPHOLD).

Luhabe[5] and notably Lyndall Shope-Mafole[6] who 'was an important player in the regulatory discussion that led to the government approval for the listing' and Andile Ngcaba[7] amongst other politically connected individuals (Gedye and Letsoalo, 2009). Telkom shareholders received a considerable dividend as part of a deal when it sold Telkom's state in Vodacom to Vodafone. Another fascinating aspect of this complex deal was that Telkom shareholders, upon listing on the JSE, resulted in the shareholders, including those from Elephant Consortium, getting a cash payout and Vodacom shares (Otter, 2009).

Another way empowerment scheme that Vodacom tries to use was the Yebo Yethu which was an empowerment entity created to buy and hold shares in Vodacom Group (Yebo Yethu, n.d). As Ayanda Cawe (2020) explains, the first version of the Yebo Yethu deal was primarily about compliance with empowerments codes.

> The first Yebo Yethu deal was structured to provide marginal but not at all insignificant ownership stakes in Vodacom's South Africa business to Thebe (0.84 per cent); Royal Bafokeng (1.9 per cent); and the Yebo Yethu Company (3.44 per cent) (holding the stakes of the black public and Vodacom's employee stock ownership programme (ESOP). The remaining over 90 per cent in equity was retained by the Vodacom Group. (Ibid.: 251)

Due to demands for restructuring BEE and BB-BEE deals overall, Vodacom restructured its Yebu Yethu, which aligns with Vodacom's value positioning in the country. Matimba Mbungela, the chief human resources officer at Vodacom Group, explaining the company's positioning shows that it views itself as a transformative force which permeates all aspects of the company's operations:

> Throughout our history, we have embraced our role as a change agent by democratising access to mobile services. Through our franchises and community payphones, we have transformed the lives of entrepreneurs and are now empowering customers through data connectivity. Over and above

[5] Well-known South African business woman and former chair of Vodacom, she is married to the one of the founders and deputy president of Cope.

[6] Former Department of Communications director-general (2004–9) and founding member of COPE.

[7] Former Department of Communications director-general (1995–2003) and led BB-BEE Consortium Thintana (the former strategic equity partner to Telkom, comprising SBC Communications and Telekom Malaysia).

> the economic impact we've had and contribution we have made to SA, we have also embraced our role as a corporate citizen to contribute to gender parity. As a company with local roots, we are committed to realising the ideals of BBBEE and Employment Equity Acts. We fully embrace transformation and its ideals which, among other things, aim to provide women with equal work opportunities so they can contribute meaningfully in the mainstream economy. (Mbungela, 2016)

The restructuring of the Yebo Yethu deal was part of a larger strategy around Vodacom's transformation agenda, which meant that Yebo Yethu moved from Vodacom South Africa to Vodacom Group.

> Vodacom will increase its ownership of Vodacom South Africa from 93.75% to 100% and issue new Vodacom Group shares to its BEE investors, diluting existing Vodacom Group shareholders proportionately. Consequently, Vodafone's shareholding in Vodacom is expected to fall from 64.5% to approximately 60.5%. (Vodafone, 2018)

The change in structure helped Vodacom meet the new demands under the BB-BBEE charter. Shameel Joosub, CEO of Vodacom Group, in response to Vodacom achieving the highest Level-1 contributor towards BB-BBEE for the fourth consecutive year (see Table 4), notes that

Table 4 Change of BEE Points Score over Three Financial Years

Scoring element	Target points	Points 2022	Points 2021	Points 2020
Ownership	25	25.00	23.23	22.75
Management control	23	15.99	18.23	15.90
Board representation	8	6.83	6.83	4.67
Top management representation	5	2.73	4.90	4.82
Employment equity	10	6.44	6.50	6.41
Skills development	20	23.04	21.97	20.90
Enterprise and supplier development	50	48.55	47.98	49.29
Procurement	25	23.07	22.05	21.29
Supplier development	10	8.48	8.93	11.00
Enterprise development	15	17.00	17.00	17.00
Socio-economic development	12	12.00	12.00	12.00

Source: Adapted for Vodacom (2022: 74).

> as a company with deep local roots, we fully embrace transformation and its ideals which, among other things, aim to provide women with equal work opportunities so they can contribute meaningfully in the mainstream economy. (Vodacom Group, 2022)

Vodacom S B-BBEE verification is based on the Amended ICT Codes of Good Practice for Broad Based Black Economic Empowerment Gazetted on 7 November 2016.[8]

A Rhodes redux? Realizing from Cape to Cairo

South African corporate expansion into the rest of Africa and beyond is a common strategy for many MNOs (see Chapter 3). As Max Dike and Elizabeth Rose (2017) explain, MNOs are motivated to expand beyond their borders for several reasons. These include the desire to seize opportunities in investment destinations and to establish a first-mover advantage. Additionally, factors such as regulatory ease and existing alliances can make expansion into certain regions more accessible and desirable, allowing MNOs to take advantage of the economic diplomacy and political goodwill the state has garnered. Post-apartheid South Africa has managed to shed its image of being a pariah state to one that is a transformative agent for African development and an ideological driver of African integration. It is against this context that MNCs, particularly MNOs, have been able to expand.

Empirical research conducted by John Luiz and Henry Stephen (2012: 627–34) has revealed that the primary motivations for South African telecommunications firms to invest in other parts of the continent are market-seeking and strategic in nature. These markets, characterized by low penetration rates, are ripe for growth and offer the added advantage of rapid network infrastructure and service deployment. South African MNOs are familiar with working in un/under-developed regulatory environments, which has helped them internationalize in developing countries.

[8] These are ownership, management control, employment equity, skills development, preferential procurement, supplier development, enterprise development and socio-economic development. Vodacom posted significant improvements on all seven components.

As Luiz and Stephen explain, 'The liability of being from the relatively small emerging market of South Africa and operating in a country of institutional uncertainty has allowed South African companies to internationalize into markets that other MNEs [MNCs] will not venture into, thus turning a liability into an advantage' (Ibid: 634). Therefore expanding, and using a South African intermediary as a springboard or gateway, is very attractive for many MNCs (Oguttu, 2011; Scholvin and Draper, 2012; Miti and Kilambo, 2012).

The fluctuating majority ownership of Vodacom raises questions about whether it is truly a South African company or simply uses its South African structure as an intermediary to expand into other African countries. In its integrated report, Vodacom states:

> We are clear in our ambition to be the market leader in the countries where we operate across Africa. For Vodacom, this reflects the quality and depth of service we can offer in each market, rather than the size of our footprint. In this way, we concentrate our efforts and investment on meaningful innovation that helps to unlock each country's potential for economic growth. (Vodacom, 2022: 38)

From its stated vision to its strategic acquisitions, it is clear that Vodacom intends to maintain its dominant market share and has also managed to achieve the dream of connecting from Cape to Cairo through a series of strategic purchases. In 2008, Vodacom acquired Gateway Telecommunications, which provides various communication services throughout the continent, including cellular backhaul and e-commerce services. In May 2019, Vodacom signalled a move towards frontier technologies by acquiring global internet of things software and hardware provider IoT.nxt. In the same year, Vodacom purchased Safaricom and the following year acquired M-Pesa. Finally, in December 2022, Vodacom completed the link by acquiring Vodafone Egypt.

Conclusion

In the twenty-first century, MNOs have evolved beyond their traditional role as network infrastructure providers and mobile telephony services. They now offer various services, including banking and other value-added offerings.

With the ongoing convergence of technology, the basket of services MNOs provides is set to expand even further.

However, despite over two decades of operation, MNOs still face the fundamental challenge of providing services to economically and regionally marginalized populations. In the South African context, this population is often Black. Considerable expectations for universal access and service provision accompanied the emergence of Vodacom and MTN. Yet, the very issuance of their operating licenses was mired in the day's politics.

The outgoing NP government sought to control the process and prevent the ANC from having a say in mobile telephony. They curated the entire process, from the partnerships that would ultimately dictate the terms of engagement to the international partnerships and network standards that would establish mobile telephony in the country. To a large degree, these political considerations continue to shape the provision of mobile services in South Africa.

Interestingly, the Multiparty Implementation Agreement in 1993 – which was between the licensees, the minister of Post and Telecommunications, and the post-master general and excluded the ANC – already largely racialized the rollout of the mobile telephony services before it even started. Both had to focus on urban areas (which were primarily 'white') 'with Vodacom targeting coverage along the major highways, and MTN including a significant number of "black" townships in its plans' (Lewis, 2018: 151).

The ANC government recognized the critical role of telecommunications provision in socio-economic development and prioritized it in policymaking through the RDP. The Draft ANC policy for Equity and Efficiency in the Telecommunications Sector was more detailed and comprehensive than the RDP with regard to telecommunications. It aims to marry universal access goals and close the racial divide in telecommunications provision.

The ANC's vision for transformation in the telecommunications sector included increased Black corporate ownership and changes to procurement policies to include Black businesses. Against this backdrop, we see the emergence and growth of Vodacom. Backed by its majority shareholder Vodafone, Vodacom had access to superior technology and experience building the first mobile network in the territory. It also had the political clout that came with partnering with Telkom.

However, as South African ownership in Vodacom dilutes, questions arise about whether it is a South African company or merely an intermediary of Vodafone. This question becomes even more critical as the company expands its presence across the African continent. Is Vodacom using South Africa as a gateway to the rest of the continent?

Ultimately, the broader lessons the Vodacom case speaks to are the complex relationships that the South African business environment births. The need for redress, service provision, but also questions around identity. What is a South African company? What is the implication of being South African for expansion? The Vodacom case shows that the answers to those questions are nuanced – and the undercurrent is still racialized.

6

MTN

Since its creation in 1993, MTN became the leading mobile operator in over twenty-one markets in Africa and the Middle East. As a relatively new sector, the rules of engagement are still being written. MTN and Vodacom were both created during South Africa's democratic transition. The story of the MTN group (also referred to as MTN) is undoubtedly impressive; an empowerment company that has become a pan-African mobile telephony behemoth in just over twenty years. Despite its success, its relationship with the State and the ruling party has often appeared to be patchy. Reasons include the reputational risk of MTN's sketchy investments outside of the continent; in 2020, the company began disinvesting from the Middle East to grow its footprint in Africa.

MTN is in a unique position because many of its current, and historical key shareholders and board members, have had direct or indirect relations with the ruling party. The creation of MTN was born of the desire for involvement of Black South Africans in the telephony space. Important ANC figures, such as Cyril Ramaphosa – who would later become the president of South Africa[1] – former president Thabo Mbeki, Irene Charnely, Mcesbisi Jonas and Nthato Motlana, have all been involved with the company in varying degrees. However, the close relationship with the ruling ANC has not helped the company to the extent expected, as the State has not come out aggressively to assist the company outside of South Africa's borders. Nonetheless, its level of power projection, particularly outside of the continent, necessitates analysis. MTN's proud vision was 'to lead the delivery of a bold, new Digital World to our customers' (MTN Group Limited, 2015). It appears that it has been

[1] He divested from his stakes in MTN prior to taking up office.

successful in achieving this vision. The MTN group has been one of the most prolific actors in the mobile telephony space. It is also increasingly a player in providing ICTs on a number of platforms and has proven itself adept at adapting to local conditions and challenges.

The underlying question of this chapter is whether or not MTN has leveraged its history, and political relationships, to further its corporate objectives; also, whether the South African government has leveraged the success of MTN to improve its standing or perception of power on the continent.

In a number of media outlets on the continent, MTN has been accused of at times displaying a propensity towards patronage.

History of MTN

Telephony in South Africa before the democracy was not evenly spread. Very few Black communities had access to telephone services, even though they made up most of the population. There was a realization within the incoming ANC government that additional telephony capacity and infrastructure were needed. As a result, the entire process of issuing the licenses and even pursuing mobile telephony as a viable solution to the dearth of communications infrastructure was highly political. The ultimate relationship between MNOs and the State has become convoluted – despite the regulatory framework that has ultimately developed. MTN provides the State with high amounts of tax remittances and increases South Africa's profile of South Africa. Nevertheless, there is sometimes anger at the lack of state control of MNCs.

The license for MTN was one of the outcomes of the multiparty negotiations during South Africa's democratic transition (Horwitz, 2001b). Nicholas Sullivan, tracing the development of the MTN group, reports that, despite opposition, some lobbied for granting a second mobile license within the ANC. Nthato Motlana, the chairman of the empowerment group New Africa Investment Limited (NAIL), spent nine months appealing to the party (Sullivan, 2007: 100).

Before the issuing of the license, plans were already in place between established white business and key players within the ANC to bid for the license. At the time this was not surprising, as key business leaders had already started engaging with the ANC on the future of a transformed South Africa. Consequently, many financial relationships were formed (Adam, Slabbert, and Moodley, 1997; Handley, 2008; Horwitz, 2001a; Naidu, 2016). The burgeoning Black business class needed the capital and expertise from established white, English and Afrikaans businesses if the ANC's plan to get Black Africans involved in the economy was to be successful.

MTN's initial shareholding was quite complex; in 1991 a multi-layered consortium, that included M-Net,[2] Cable and Wireless,[3] Transtel[4] and National African communications[5] (NAFTEL), created M-Cell (the holding company of MTN). Through NAFTEL and future increase in equity in M-Cell, NAIL would own 20 per cent of the company while the South African Black Taxi Association and SACTWU each had a 5 per cent holding (Sullivan, 2007: 110).

The creation of MTN and MTN group was not only fraught with tensions from the outset, but it also highlighted great incongruities of the ideology of some of its shareholders, most notably COSATU. COSATU was one of the main agitators, outside of and within the ANC tripartite alliance, against issuing mobile licenses. The concerns were that it would cause job losses for Telkom workers and cell phones were essentially an elitist indulgence.[6] However, financing from COSATU and its affiliates helped create MTN indirectly through the National Empowerment Consortium. Tomaselli confirms the paradox inherent in union-generated Black investment capital in the 1990s. A considerable part of Black investment capital was generated from Black unions' pension and provident funds that amounted to 'R20 billion'. The ANC government used these funds for its privatization initiatives, even though COSATU vehemently opposed privatization (Tomaselli, 2000: 257). Nonetheless, it willingly allowed this money to be invested in a process it opposed (Ibid.). The State could do this as it was in control of the funds.

In 1993, the process of bidding for the MTN licence began (Ndemo and Mureithi, 2015: 187). Between 1994 and 1995, M-Cell held 25 per cent of MTN holdings and 60 per cent of M-Tel (Goldstuck, 2006; see Figures 3 and 4).

[2] A South African pay television company established by Naspers in 1986; Naspers was created by the NP in the early twentieth century.

[3] UK based company. It eventually helped build the physical infrastructure of the MTN.

[4] The communications arm of parastatal Transnet.

[5] Part of NAIL.

[6] These issues were referred to in the previous chapter.

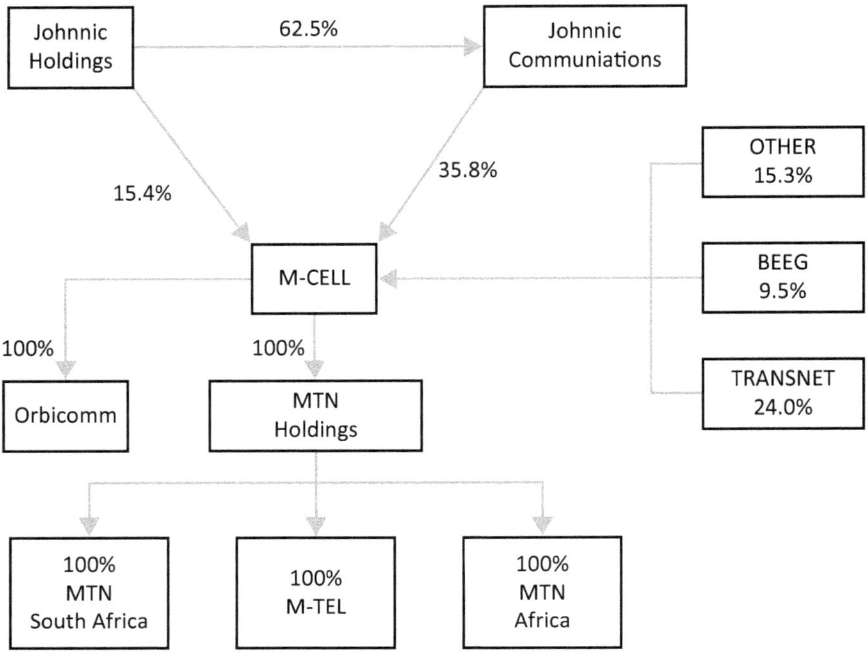

Figure 3 MTN/M-Cell ownership structure on 31 March 2001.

Source: M-Cell, 2001: 2. Edited from original relationship structure. BEEG stands for Black Economic Empowerment Group.

The latter eventually became MTN's service provider. M-Cell began as a shelf business, Investment Facility Company Two Six Five Ltd, in November 1994. This shelf company was subsequently converted into a public company and changed its name to M-Cell Ltd. M-Cell, between 1997 and 2000, increased its stake in MTN holdings and in the year 2000 MTN Group Ltd. was born.

The eventual creation of MTN group was rife with the struggle to maintain its character as an empowerment company as the company expanded. Furthermore, some of the empowerment investment vehicles with shares in MTN were fraught with internal strife (Sigonyela, 2001). The extent it has gone to, to elevate its empowerment status, signals the importance placed on it being true to its original mandate as an empowerment vehicle. MTN's South African unit, MTN SA, created one of the most extensive BBBEE deals to date (Reuters and I-Net Bridge, 2010) and thus elevated its BEE status to Level two (mybroadband staff, 2011). This means that according to South

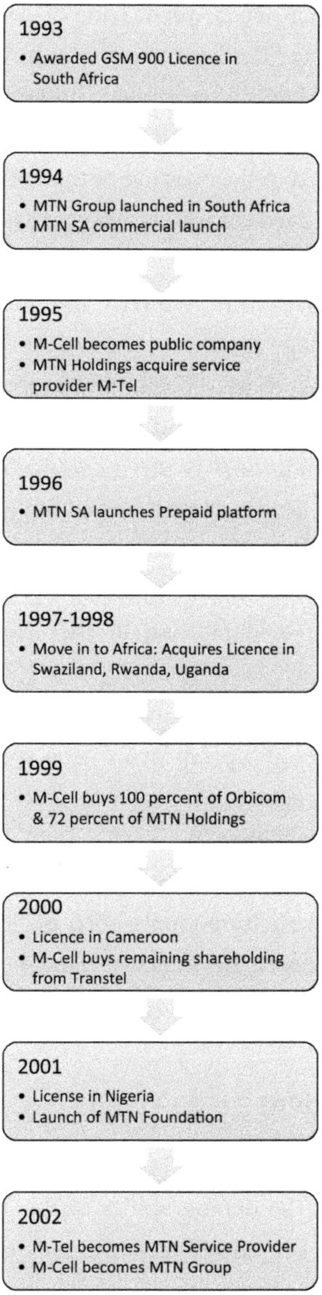

Figure 4 MTN Group founding timeline.

Source: Adapted from Goldstuck, 2006: 33–5.

Africa's BEE legislation, and consequent code of conduct, it is considered a fully empowered[7] company (The Republic of South Africa, 2003).

When South African MNOs were growing and expanding their operations outside of South Africa, another paradigmatic shift was happening. The pan-Africanist ideology that led to the creation of the OAU in 1963 was beginning to take a different form as Africans wanted a bigger say in their economic future and not only in their political future. There was more receptiveness for African multinationals in comparison to Western multinationals. Buoyed by the African Renaissance ideology and the creation of the AU, NEPAD and attempts at regionalization, MNOs such as MTN found a more welcoming environment when they began their quest to expand. In many locations they were further helped by the fact that traditional Western companies were not interested in moving into Africa at the time because they had not yet perceived its potential.

South Africa illustrated its desire to integrate into Africa politically and ideologically, and the private sector wanted to be part of this integration. Nowhere was this more evident than in the telecommunications sector, particularly by MTN.

By 1997, MTN had made its first moves in Uganda, Rwanda and Swaziland (Cairns, 2010). The MTN group report in 2009 speaks of a historical 'pre-emptive-move strategy to aggressively enter markets in Africa and the Middle East' (MTN Group Limited, 2009: 2).

MTN's expansion differs from other South African-born multinationals as it was one of the first multinationals born of South Africa's democratic transition. Its history reflects the structural and political change that took place in the country and within the ruling party.

MTN's African operations

Unlike many companies at the time, MTN was emblematic of a burgeoning Black industrial elite. As the demographics of the MTN shifted from being primarily white to being primarily Black, the importance of the company as

[7] An empowered company is one that has met or surpassed the requirements of the B-BEE codes. The approach of the codes is to 'situate black economic empowerment within the context of a broader national empowerment strategy focused on historically disadvantaged people, and particularly black people, women, youth, the disabled, and rural communities' (BrandSA, 2013).

the face of the 'new' South Africa grew. The country was very willing to export this image, and the success of the MTN's proliferation in the continent was significant to the state and the country's political elite.

Neuma Grobbelaar points out that business cannot be separated from its 'sensitive political and social environment' (Grobbelaar, 2014: 46). Grobbelaar's conclusions are based on a four-year South African Institute of International Affairs (SAIIA) research study on South Africa's corporate expansion. In analysing the impact of South African businesses on the environment, some of the points that she highlights include the fact that consumer culture was often changed, the corporate governance environment was changed, the work environment was often more attractive and trade, technology transfers as well as local economic development were impacted.

MTN began its expansion by acquiring licences in the 1990s in Swaziland, Uganda, Rwanda and notably in Nigeria in the early 2000s. Each of these markets were not only strategically important for the company – and ultimately quite profitable. But, Ewan Sutherland (2014a), and others,[8] point out, the complex structure of MNOs make real ownership opaque. For companies like MTN it creates additional complications because of its exposure to political figures in other countries and is often being viewed as a proxy for South Africa.[9]

MTN's corporate strategy for Africa underscores its export of South African values and governance model. For instance, MTN places considerable emphasis on local ownership and utilizing local supply chains, which is very similar to its South African operations.

I do not address whether MTN is an actor in conflict or peace-building activities in its host economies. I focus on whether these operations contribute to South Africa's power projection.

Corporate identity: A political torque?

Projecting an image of being a responsible corporate citizen has increasingly become the norm, and sustainable development is undoubtedly a keyword in

[8] See Whalley and Curwen (2013).

[9] In response to unhappiness with South Africa, often in the aftermath of xenophobic violence in the country, MTN facilitie are often targeted (see Channels Television, 2012; Addeh, 2017; Reuters, 2019; AP, 2023).

business, both big and small, in the twenty-first century. A good corporate citizen means providing goods or services for those at the bottom of the pyramid, including corporate social responsibility and social investment. This new set of norms is reflected internationally, for example, through the UN Compact and the OECD business guidelines. However, there has been little consideration for the implications of business involvement in a government domain. This consideration goes beyond the politics of liberal versus socialist ideals.

The real issue is that the State tasks the private sector with essentially providing goods and services to the poor while at the same time helping them move out of the cycle of poverty. Both are noble and honourable endeavours, but the long-term implications of this level of responsibility are rarely considered. It must be remembered that most enterprises, especially transnational ones, are not designed with social welfare as their core competency.

Since the international system is being modified to reflect a new set of norms for business, whether mobile telephony companies are the torque or the axis must be considered. In physics and mechanics, torque is defined, in layman's terms, as 'the amount of "turning power" you have, much in the same way you turn a wrench' (Siegal, 2009). In other words, it measures the force that moves an inert object from one point to another along an axis. If we apply this principle to our query in international relations, the torque would be a measure of the political will that enforces change within the larger political economy.

Throughout its various host economies, MTN is often seen as a desirable employer.

As MTN expanded in Africa, its hiring practices evolved. It created an 'Expat Pool' that was made up of staff identified from all of its operations that it deployed rapidly to set up new operations (Maketa, 2009: 29). Its additional 'mandate' was to transfer skills and build local talent in each operation (Ibid.: 30). The hiring of majority local staff, as well as local subsidiary ownership, is the hallmark of MTN's expansion strategy. For instance, in Cameroon, the company progressively reduced the number of expatriates on its staff to five in 2007 (Chinje, 2008: 55). The MTN Group Sustainability report for 2015 shows that there was only a 0.8 per cent expatriate staff base across the group (MTN Group Limited, 2015).

The company has consistently aimed to create a local subsidiary ownership scheme in which the nationals own a considerable portion of the company. This plan does not always work for various reasons, and the result in places such as Nigeria is that the companies become majority foreign-owned. The desire to increase local ownership in Nigeria is the main driver behind the new share scheme launched on the Nigerian Stock Exchange in 2016. What does all this mean? As much as Nigeria is a financial success for MTN, continuing to foster close relations with the country politically has been a boon for South Africa. Nigeria's level of influence within west Africa and the continent gives legitimacy to South Africa's involvement and its quest for leadership.

MTN's drive for a large portion of local ownership flies in the face of what is the norm for traditional multinationals. Interestingly, the share release for local ownership resembles South Africa's BEE share schemes. Even though the process has many critics, it has successfully gotten previously disadvantaged people involved in the corporate sector. MTN Zakehele, which launched in 2010 and has been trading on its platform, was widely successful, selling at several more times than it originally paid for in 2010 (Ujuh, 2015).

The year 2015 saw the rise of Black shareholder activism, with shareholders of these BEE shares attending the recent MTN AGM to voice their concerns about how things are run at MTN. In the Sunday Times Business section, Speckman (2015) reported that a number of these concerns surrounded the implementation of the vision of BBBEE, such as the request for clarity on MTN's procurement policy of Black SMEs. As surprising as this may have been for MTN senior management, this aligns with the vision that the ruling ANC and government had for South Africa – and perhaps the rest of Africa. Although much of Africa was liberated from white minority rule in the 1960s, the process of company indigenization has not been successful. It may be argued that the indigenization of large corporations has failed in much of Africa. The perception is that MTN is transposing a South African indigenization model to its different host countries – particularly in Africa.

Arguably, South Africa's version of indigenization is BEE; when MTN operates outside its borders, it works towards localizing the value chain and keeps its expatriate staff levels low. Moreover, it works towards local share ownership rather than having the local entity primarily South African owned.

Corporate governance is another area where South African models seem to be transposed in the rest of Africa by MTN. An evident corporate governance culture has not been the norm in several African countries. Regarding corporate governance and anti-corruption stances, in its 2014 Corporate Governance report, MTN lists its efforts at group-wide anti-bribery and corruption programmes and its channels to facilitate whistle-blowing (MTN Group, 2014: 16).

Private enterprise and the ever-hanging cloud of political patronage: Corruption or relationship building?

Expanding to a new territory requires a degree of relationship building with the ruling political elite of the investment destination. A complaint often levelled against African multinationals is that they entrench patronage networks and thus encourage corruption (Tangri, 1999). A counter to this argument is that a degree of political patronage is, in fact, a form of relationship building. If this counterargument is accepted, there is still a whisper-thin line before patronage crosses over to outright corruption. Lyle White explains that most business in Africa is done with governments, as most projects are State driven. Foreign investment is a relationship and confidence-building exercise between a company and the country it invests in (White, 2016). Adrian Meyer, quoted by Alex Weingrod, provides a pithy description of patronage: 'The transactor (patron) has the power to give some benefit which the respondent (client) desires' (Weingrod, 1968). With the government being the patron and the private sector being the client, the playing field can become murky, and fairness disappears. Transparency International defines *political corruption* as 'a manipulation of policies, institutions and rules of procedure in the allocation of resources and financing by political decision-makers, who abuse their position to sustain their power, status and wealth' (Transparency International, 2015).

MTN has not been immune to accusations of benefiting from patron-client privilege in its host economies. Moreover, the close ties between MTN and the ruling party create the impression that MTN has been given more license to engage in questionable behaviour.

The political positioning of South Africa has given South African multinationals the ability to set up shop in the rest of the continent quickly. One of the most pervasive narratives in popular culture about multinationals relates to the big bad parasitic company. Analysts such as Ewan Sutherland (2015) contend that mobile telephony companies can penetrate the market with politicians' help and possibly greasing palms. However, the true State of affairs is complex and multi-layered. Having a high political influence makes a difference because operating on the African continent is no mean feat. Reports have been there of politicians stepping in to smooth the road for MTN. For instance, it was reported that South Africa's President Thabo Mbeki and President Thomas Boni Yayi assisted MTN Benin in brokering a deal when the company had an issue with the Benin telecoms regulator (Muller, 2007a).

There have also been many questions surrounding MTN executives and their ability to leverage their struggle relationships within the ruling ANC government. For instance, it was reported in South Africa's *Times* newspaper that Irene Charnley wrote to President Mbeki about the controversial Iran license requesting 'an intervention from your office'. She was also reported to have written to the Former Foreign Affairs director-general, Ayanda Ntsaluba, offering to arrange a meeting between President Hassan Rouhani and then president Mbeki as a 'token of our appreciation for your support as we scoured the Middle East for licensing opportunities' (Hofstatter, Rose, and Wa Afrika, 2013).

A number of the MTN board members, past and current, have, at some point or another, had ties to the ANC – most dating back to the anti-apartheid struggle. Even its chairman between 2002 and mid-2013, Cyril Ramaphosa, is a long-standing member of the upper echelons of the ruling ANC. On the surface, it appears that MTN's ability to penetrate the African markets relates to its long-standing relations with South Africa's political elite. It is uncertain whether the company would have had the same level of success without its close relationship with political power.

Having that level of apparent political backing is illustrative of MTN's pull. Dianna Games notes that MNCs often have had to use their political influence. She notes that although companies may try to bypass risky political connections and take slower official bureaucratic channels, there is a realization that 'close ties with government can reduce regulatory risk. When unreasonable or ad hoc regulations are introduced, companies at least have a channel of appeal if they have the ear of top officials' (Games, 2011: 39).

As a whole, telecommunications has grown increasingly important, and the sector's value in forwarding South Africa's foreign policy objectives is only recently becoming apparent. The focus of South African foreign policy is the deepening of SADC integration and the overall renewal of Africa. These priorities are encapsulated in the *South African Foreign Policy Discussion Document* (1995). Several other documents have repeated these foci, including DIRCO's *Medium Strategic Framework of 2009–2014*. Another key interest that has popped up in South Africa's foreign policy documentation and posturing has been the strengthening of South-South relations. In 1995, Foreign Minister Alfred Nzo stated, 'South Africa should deal with African partners as equals and avoid all hegemonic ambitions' (Department of Foreign Affairs, 1996: 16). South Africa has sought to create and maintain relationships and institutional regimes entailing its leadership. When MTN embarked on its expansion efforts on the continent compared to the maturation of South Africa's diplomacy organism is considered, it is hard not to imagine that there was not some form of grand design. Though, it is more likely that MTN was just fast in its response to political developments on the continent.

In 1997, MTN expanded to Uganda, Rwanda, and Swaziland. Those territories were very curious choices for initial expansion. Although Swaziland is a tiny country surrounded by South Africa, the choice was curious from a market point of view. Swaziland can cause serious reputational risk to the company because the country's corporate governance standards are quite different from many other locations. It was a partnership between the Swazi Government, MTN, and the Royal Family. The king holds 10 per cent of the shares in MTN (Swazi Media Commentary, 2015), which makes any competing network opening unlikely. The fact that the king's eldest daughter was also appointed to the Board of Directors of the local MTN subsidiary (Lukhele, 2012) illustrates a high level of local political investment in the company. In his article, *Telecommunications in the Kingdom of Swaziland*, Ewan Sutherland (2014b) bemoaned the country's governance system that prioritizes the sovereign's interest. As much as poor corporate governance practices are a reputational risk to MTN, it also is a reputational risk to South Africa.

The political climate in Swaziland has been a cauldron threatening to bubble over for many years. This was patently obvious during the intensive strikes in 1994, 1995, 1996 and 1997.

Salmond observes that in addition to meetings between South African President Nelson Mandela and Swaziland's King Mswati III, the business sector got involved in easing political tensions (Salmond, 1997). Swaziland has always been dependent on South Africa, both politically and economically. It is well-known that South African businesses used Swaziland to evade sanctions. It is no surprise then that the ANC used the same territory that caused it much trouble during the fight against apartheid in the same way.

Langalanga has argued that South Africa, since 1994, has influenced Swaziland's domestic political atmosphere through institutions such as SACU and SADC, as well as non-state actors. It is argued that 'the presence of South African companies and State-owned enterprises could provide much-needed political and economic leverage to ensure greater tolerance of diverging views. The threat of their withdrawal would directly affect the Swazi establishment. However, South Africa either pretends to be unaware of its economic clout or decides to turn a blind eye, further eroding the moral capital built up during and after its transition to democracy' (Langalanga, 2014: 4).

The impression is that the South African State is willing to ignore or even prop up Swaziland (through SACU and the proliferation of South African companies) to entrench its regional dominance. However, it must be recognized that the failure of any of the small states in SADC also serves as a significant risk to South Africa that could be a beacon for fleeing citizens.

Relations with Uganda and Rwanda allow the South African government to influence Central and East Africa. Rwanda, a French-speaking country, as well as a former German then a Belgian colony, presents an opportunity to break the influence of France in francophone Africa. South African influence has been severely constrained in francophone countries because of language barriers and the influence France exerts in former French Africa. South Africa's most viable entry point into francophone Africa is Rwanda, owing to its tenuous relationship with France; France is accused of supporting the perpetrators of genocide by the Rwandese (Rubin and de la Baume, 2014).

Indeed, even before the lead to the 1994 genocide, the French involvement in Rwandese political affairs was disconcerting. Ironically, in the aftermath of the genocide, Western and African Anglophone countries started to make headway in Rwanda, particularly South Africa.

Since its advancement into Rwanda, South Africa has intensified its relationship with the country. This is evident in the increasing level of foreign direct investment from South Africa in the reconstruction of Rwanda. In particular, a key area for the Rwandese has been ICT and mobile telephony. They had struggled to get foreign interest in helping bring mobile telephony to the country. In partnership with the ruling Rwanda Patriotic Front (RPF), South Africa's MTN helped leapfrog the country's ICT development on the continent. The RPF and its holding company Crystal Ventures Limited (previously Tri-Star) provided much of the capital to establish MTN Rwanda. It was one of the critical initial stakeholders. The result of the risk that Tri-Star took was that it fast-forwarded Rwanda into mobile telephony and provided 'spill-over benefits "for the entire sector, one of them being the experience gained by entrepreneurs dealing with MTN" on behalf of the government' (Booth and Golooba-Mutebi, 2012: 398–9).

From an economic point of view, Rwanda should not necessarily have been MTN's first choice of investment destination. Its relatively small population and decimated infrastructure made it unattractive to other players; one has to ask why MTN felt that it was worth the risk to partner with Crystal Ventures/Tri-Star leading to the launch of MTN Rwanda in 1998. However, to assume that the corporate expansion deepened relations between South Africa and Rwanda would be incorrect.

The diplomatic spat between Rwanda and South Africa, which emanated from the expulsion of Rwandese diplomats from South Africa after the assassination of Patrick Karegeya in 2014 on South African soil, is evidence of a tricky relationship. Karegeya, a former intelligence operative, was an asylum seeker in South Africa (Allison, 2014). Both countries have insisted that they are intent on smoothing relations between their countries (Kotch, 2014). Part of the issues between Rwanda and South Africa may have to do with the leadership of the Great Lakes Region and overall primacy in Africa. For peacekeeping missions, Rwanda is one of the top contributors. Its involvement in security and economic matters in the Great Lakes region makes it a key player.

Another influential country in the Great Lakes region is Uganda. Unlike Rwanda, Uganda already had mobile phone services in the form of Celtel. When MTN entered the market, it quickly dominated it. In this locale, its model includes mobile line services, public payphones, and internet services. Uganda has always had a very cordial relationship with South Africa. Economically, there has been a long-standing bilateral relationship: in 1997, they signed the Convention on the Avoidance of Double Taxation and the Prevention of Fiscal Evasion; in 2000, they signed an Agreement on the Promotion and Protection of Investments; and, in 2002 launched the Joint Commission of Cooperation. Further entrenching the relationship between the two countries through South Africa's private sector also makes sense.

The lawsuit by a competing company, Turkcell, against MTN in issuing the Iranian GSM license fell through. MTN was found to have won the license legitimately, but the process leading to their victory did have its casualties and brought several things to the fore (IT News Africa Staff, 2014). A KPMG forensic report noted that in the run-up to the license issuing, both Iranian officials and officials of the South African foreign services were treated to luxurious gifts (Hofstatter, Rose, and Wa Afrika, 2013).

It was also reported that MTN sponsored Iran's deputy trade minister's trip to South Africa, where he attended a dinner hosted by the Defence Department with then MTN chairman Cyril Ramaphosa (Hofstatter, Rose, and Wa Afrika, 2013). There is no proof that this level of lobbying was corrupt or over what is typically done in practice, but it does raise many flags. Moreover, it also brings into question the nature of the lobbying for MTN to get a license in other countries. In the aftermath of the Iran scandal, the MTN finance chief, Nazir Patel, resigned amidst allegations that he committed several governance breaches in the Iran debacle (Marais and Rose, 2013).

Other than various trade promotion strategies undertaken by the government, and the various agreements and regimes that are entered into, there is no explicit instruction that South African multinational companies enter other territories to expand the influence of South Africa. South African companies expanding onto the rest of the continent do so with very little help from the South African government. Nevertheless, the fact that the government regularly enters into relationships with and takes a stance that is beneficial to the private sector, and at the same time the deepening of ties beneficial to South Africa, cannot be a coincidence. There is an interesting relationship between South Africa's corporate engagements and the State's foreign policy decision making.

An interesting case is that of Nigeria, MTN's largest market. During President Obasanjo's administration, a decision was taken to open Nigeria's market to mobile operators. During this period, Nigeria and South Africa had very cordial relations, coinciding with the time that NEPAD was promoted as a pan-African development agenda. Interestingly, the NEPAD document recognized the value of ICTs and the promotion of African business. MTN emerged in this environment, and a close examination of its expansion efforts reveals a parallel to the relationship-building exercises undertaken by the State. MTN moved into the Nigerian market when Nigeria and post-apartheid South Africa began establishing ties officially.

Nigeria and South Africa have had the continent as one of the pillars of their respective foreign policies. For instance, as much as South Africa has been a critical driver of NEPAD, Nigeria has also supported the call for its development. Both former president Obasanjo and former president Mbeki campaigned relentlessly in the name of NEPAD and the African Renaissance.

There has been a myriad of affronts in this contentious relationship. Economically it is clear that there is an asymmetrical relationship. This is due mainly to South Africa's infrastructure, technical know-how and easy access to capital. When South Africa and Nigeria signed bilateral agreements in 1999, the door was opened for big players – such as MTN.

South African corporate engagement on the continent predates the development of NEPAD or even Mbeki's African Renaissance ideology. However, NEPAD was the first African-owned developmental plan. Companies such as MTN took the opportunity to interweave their corporate narrative with that of NEPAD and the African Renaissance. Its presentation at the 2002 Business-NEPAD conference highlights MTN's contribution to meeting NEPADs ICT priorities. In a journal article born from the MTN presentation, Yvonne Muthien and Meshack Khosa further confirm MTN's commitment to NEPAD and its belief in contributing to the African Renaissance (Muthien and Khosa, 2002).

NEPAD was launched to the excitement of the business community; the only concerns were tax burdens (e.g. double taxation) and investment controls

(Esipisu, 2002). Many of these concerns were addressed in treasury's Gateway Initiative.

In an example from Nigeria, when MTN was awarded its license in 2001, its investment in the country was the largest it had ever made outside South Africa (Onuha, 2008). Undoubtedly, NEPAD assisted in making this a reality (before this date, MTN had attempted twice to penetrate the Nigerian market). Thanks to NEPAD, South African SOEs such as the IDC and many other private telecommunications companies have made significant inroads into Nigeria.

Due to MTN Nigeria's value, Cyril Ramaphosa's investment firm, Shanduka group's Mauritius unit, bought a US$335 million stake in MTN Nigeria (Nyambura-Mwaura, 2012). This purchase gave MTN group an estimated 78.8 per cent stake in MTN Nigeria and a considerable amount added to the remaining 21.2 per cent belonging to Shanduka. Pascal Dozie, an MTN Nigeria shareholder and an MTN Foundation board member, explained that the initial reluctance of Nigerian investors explained the share structure. The original plan was for Nigerian interest in the company to be up to 40 per cent (Nurudeen, 2015).

Ultimately, this all comes down to how MTN and other South African multinationals view their role in the socio-economy of their host nations. In other words, when they develop, they develop the narrative of their corporate strategies and what values underpin their operations. In the case of MTN, the values that underpin their operations sound very similar to that of the South African government (discussed in Chapter 4). In a meeting with MTN's Chris Maroleng, he notes that MTN is not a 'philanthropist' but does see the more considerable value of socially empowering its host economies (Maroleng, 2016).

The expansion strategies of the MTN group mirror the clear vision and strategy of South Africa's government: to unify the Global South and 'lead' the development process on the continent. The South African government is present and influential in several fora by a mixture of design and the advantage bestowed upon it owing to its unique history. Despite its history, the apparent imprint of South African values on MTN's corporate culture is not surprising. Other than having very politically connected people at the helm of the business at its genesis, the backgrounds of the people who assist in stakeholder relations are also very telling. For instance, the background of the people in group executive for corporate affairs positions indicates how important external stakeholder relations are to them. An example is Yvonne Muthien, a former group executive for corporate affairs who had served under former president Mandela in the Public Service Commission and both presidents Mbeki and Mandela as the president's Advisory Council for National Order chairperson. During her tenure at MTN, it was the first telecommunications company to issue a sustainability report. They clearly state their socio-economic value proposition as:

> MTN, we believe that working on the continent should be as much about the development of Africa as it is for business' sake. (M-Cell/MTN Group Corporate Affairs, 2002)

To ensure that their relations with DIRCO do not become superficial, MTN hosts several events, much like the 2011 business lunch in which it hosted all the ambassadors of its various operating companies; this lunch coincided with the annual Heads of Mission Conference (Manco Voice, 2011). Maroleng confirms that MTN's relationship with DIRCO is good, and 'they help to facilitate engagement, especially at initial stages' in new operations. He does concede, however, that DIRCO's engagement is limited because the number of South African staff in their various operations is limited (Maroleng, 2016).

Despite MTN's strategy of localization in its various operating companies, the reality is that it is still ultimately viewed as a South African company by the South African government. This was made glaringly obvious when the attempts to tie up Indian-owned Bharti Airtel (Bharti) with MTN failed dismally owing to South African authorities. Bharti had tried to partner with the MTN group a number of times, and in a media statement announcing the deal, Bharti stated that:

> this opportunity also represents a first of its kind in developing an Indian-African initiative that would serve as a shining example of South-South cooperation. (Bharti Airtel Ltd, 2009)

The *Wall Street Journal* points out that if the transaction had been successful, it would have created 'the third-biggest wireless carrier after China Mobile Ltd. and the U.K.'s Vodafone Group PLC' (Sharma and Bellman, 2009).

Outwardly it would appear that South Africa would be interested in having further commercial ties with India owing to their relationship in BRICS and the India-Brazil-South African (IBSA) Dialogue Forum. However, there were severe concerns on the part of the South African government regarding the constitution of the deal, which would have changed the character of MTN. The official statement of South Africa's treasury maintained that:

> the South African government believes that the structuring of such partnerships is best left to the companies themselves, who must make their decisions on commercial grounds. However, when companies structure their relationships outside the current exchange control regulatory framework for such transactions, they require the approval of the Minister of Finance. (National Treasury, 2009)

The minister of communications at the time, Siphiwe Nyanda, stated:

> It would be sad if we saw this entity move into the hands and management of foreign nationals. (Tarrant, 2009)

His spokesperson, Tiyani Rikhotso, even more bluntly revealed South Africa's view of the company: 'We obviously value MTN as a South African company; we want it to retain that character' (Business Standard Online, 2009). He maintained that the company had benefited from 'trade and bilateral agreements negotiated by South Africa's government' (Ibid.). This illustrates a complex interplay of interests in the relationship between South Africa and its telecommunications companies, specifically MTN. On the one hand, South Africa wants the company to be successful and it acts as a strategic asset. On the other hand, it has to walk a fine line lest it be accused of protectionism that goes against its WTO obligations. On MTN's part, despite such a deal being attractive, it needed to be careful in retaining its African character.

The economic dimension of South Africa's foreign policy engagements had benefited MTN, and MTN in return ensured that South African values had been woven into its corporate strategy. This can be ascribed to two factors: first, according to Minister Nyanda, its African expansion was made possible, 'through government help' (Reuters, 2009); second, MTN's licencing conditions in South Africa entail that any possible change of equity is subject to a public hearing process (Staff Reporter, 2009). The fact that MTN would have been subsumed by Bharti was clearly an issue for the South African government; the former treasury director-general Lesetja Kganyago stated that 'the decision will be taken by the executive … A transaction of this magnitude is not a decision you just take on your own' (Business Standard). A number of merger talks between MTN and Reliance Communications, another Indian telecommunications behemoth, also failed (Anand, 2008; Spillane, Alexander, and Antony, 2013).

MTN is clearly not afraid of risk and has entered deals that have given it a foothold in North Africa and the Middle East, such as Syria and Iran. This aligns with South Africa's attempt to deepen diplomatic ties with the Middle East and its larger strategy to align with the Global South. In a lecture series for the Emirates Centre for Strategic Studies and Research, Chris Landsberg underscores South Africa's increasing political interest in the Middle East and increasing economic engagement from 1999 onwards (Landsberg, 2007). At a 2011 meeting held at SAIIA, then deputy minister of international relations and cooperation, Ebrahim I. Ebrahim stressed the continuing importance of the Middle East and North Africa (MENA) for the South African government:

> We have followed the political developments in the Middle East and North Africa with interest because there is an inextricable link between our developmental aspirations, and genuine stability in these two regions. To us, these regions remain extremely pivotal to the realisation of our trade policies and strategies. (Ebrahim, 2011)

South Africa's reorientation to countries of the Global South would make it very concerned with relations in the developments in the MENA region. South Africa has even increased its participation in the Middle East peace process. In 2016, minister of international relations and cooperation Maite Nkoana-Mashabane participated in the Ministerial Meeting on the Middle East peace process in Paris (Department of International Relations and Cooperation [DIRCO], 2016a). In this context, MTN has increased its engagements in the MENA region despite security issues.

MTN has an office in Dubai that serves as its procurement hub for the MENA region as well as for tax efficiency (Maroleng, 2016). The company operates in volatile environments such as Syria, Yemen, Afghanistan, Sudan and South Sudan; it even has operations in Iran. Owing to the environment, MTN has struggled to repatriate its funds in these environments but has chosen to continue its operations there. It has to be kept in mind that, even at the beginning of its expansion in the late 1990s, MTN focused on underserved markets with massive populations and being politically ignored (which is more the case in North Africa than the Middle East). In an interview in PricewaterhouseCooper's Communication Review, MTN Côte d'Ivoire's CEO Wim Vanhelleputte remarked:

> Another big success factor for MTN, whether in Côte d'Ivoire or the rest of Africa, has been our brand and positioning the company as being the true African success story. If we didn't have that story as a core part of our brand, our success would be different. I'm not saying that we wouldn't be successful but that our success would be different. If tomorrow MTN were bought by Vodafone and it rebranded MTN as Vodafone, I think we would be losing a certain value, a certain extra that we have. (Dunsby and Chochan, 2012: 54)

The telling phrase is 'African Success Story'. This phrase is very similar to the external image linked to South Africa. Thanks in part to its relatively successful transition from apartheid, and its stability compared to other African territories, South Africa is considered, internally and by outsiders, as a success story. Whether this belief is correct is subject to debate. Its successful involvement in the MENA region is just a continuation of this success story which has implications for South Africa. In 2015, the Nigerian Communications Commission (NCC) issued MTN Nigeria with a fine of US$5.2 billion for failing to disconnect unregistered SIM cards by mid-2015. According to Chris Maroleng, the MTN group executive for corporate affairs at the time of the interview, MTN had actually reported itself to the NCC and had removed a considerable amount of these unregistered SIM cards off the network at the time of reporting; these unregistered cards accounted for less than 1 per cent of MTN's Nigerian subscription base (Maroleng, 2016). When the fine was issued, the recently elected Nigerian president, Muhammadu Buhari, had replaced the NCC. As a consequence, the NCC was without a board, and by NCC regulations, could not enforce the fine without the involvement of the presidency. This fine can be viewed in two parts: first, as a message to the South African government, and second, it could be a sign both internally and externally that the new government was trying to assert its authority.

At the same time MTN Nigeria was dealing with the NCC fine, many MTN subsidiaries across Africa were also embroiled in legal tussles. For instance,

MTN was ordered by Uganda's high court to pay US$662000 in an anti-competitive behaviour suit (van Zyl, 2015a); and in Liberia, Lone Star Cell/ MTN is appealing against a US$19.2 million award to the Liberia Revenue Authority (Sieh, 2016).

These legal debacles appear to be simply state's fulfilling their mandates in their territory. But, on further examination, it appears to be a case of MTN being used as a tool to send a message. A number of reasons were touted in Nigerian media as the reason that the MTN fine was harsh – security being one of the main ones. Before the fine was issued, an Ericsson Nigeria engineer was found to be installing transmitters on MTN base stations. MTN owns 12,000 base stations in Nigeria. The transmitters illegally transmitted the Radio Biafra signal on behalf of the separatist group Indigenous People of Biafra (IPOB) (Ubabukoh, 2015). President Buhari, at a joint press conference with South African president Jacob Zuma blamed MTN for the prolific attacks by the terrorist organization, Boko Haram. He reportedly said that:

> unregistered [sim cards] are being used by terrorists and between 2009 and today, at least 10,000 Nigerians were killed by Boko Haram … Other mobile phone operators complied with a mid-2015 deadline to register all sim cards … unfortunately, MTN was very slow and contributed to the casualties. (BBC News Staff, 2016)

Buhari's statement appears to be putting all the blame for Nigeria's security situation on MTN. MTN maintains that it has always responded positively to Nigeria's security services' legal requests for mobile intercepts (Maroleng, 2016). Dianna Games provides a helpful analysis. Games contends there are several reasons behind the excessive nature of the fine: the impression that MTN is making a lot of money out of Nigeria and it should recoup some of it, the relationship between security and the unregistered mobile phone connections, MTN not meticulously observing Nigeria's laws, as well as the desire to reduce Nigeria's considerable budget deficit (Games, 2015).

As much as MTN was wrong in failing to meet the deadlines given by the Nigerian authorities, the punishment has to fit the crime. Operating differently to most external multinationals, a considerable portion of MTN's funding is derived from Nigerian banks. This ensures that the local banking industry is developed and the local ties to the company are entrenched. The fact that MTN was fined for not following the laws and regulations of Nigeria, by terminating unregistered SIM cards is not in contention; what is in contention is the size of the fine. The sheer size of the fine gives the impression that the government, through the regulator, was trying to send a message to the multinationals operating in the country rather than purely punishing MTN.

The reason that the size of the fine can be interpreted as a message is that, as previously mentioned, MTN, for the most part, has operated very differently from other MNCs. A significant complaint levelled against MNCs, in general, is that they don't use locally available channels such as banking and loan facilities. Making use of these services, especially loan facilities, of a local bank injects more cash into the local system and in turn enables the bank to provide better services. Games observes that some companies shy away from local finance because of the high interest rate (Games, 2011: 20). By gaining finance locally, MTN's loan repayments inject money into the local economy.

MTN Nigeria has a long history of using locally available financing mechanisms. For instance, in 2010, it took out additional loans with a consortium of fifteen local banks that amounted to N318 billion (Ifeakandu, 2010), which converted to just over US$2.10 billion. It further took a dollar loan amounting to US$450 million from two foreign banks, Germany's KfW Ipex and Industrial and Commercial Bank of China, to cover the hardware cost. The Nigerian banks that form part of the Nigerian Consortium are Access Bank, Afribank Nigeria, Bank PHB, Citibank Nigeria Limited, Diamond Bank, Ecobank Nigeria, First City Monument Bank, Fidelity Bank, First Bank of Nigeria, Guaranty Trust Bank (GTB), Stanbic IBTC Bank, Standard Chartered, Bank Nigeria, Union Bank of Nigeria, United Bank of Africa and Zenith Bank (I-Net Bridge, 2010). In 2013, it further borrowed the equivalent of US$3 billion through GTB as well as from foreign banks such as Citigroup, Standard Chartered, Industrial and Commercial Bank of China, China Development Bank and China Construction Bank (Meineke, 2013).

This fact is important because all these Nigerian banks are the primary retail banks in Nigeria and are the lifeblood of the Nigerian economy. The failure of MTN to service these loans would devastate the Nigerian economy. Furthermore, the fact that MTN chose to take Naira-denominated loans, and pay it back as such, removes much of the accusation that it is taking away much-needed foreign currency from the Nigerian reserves.

MTN Nigeria has further stimulated the local economy across various platforms. For example, in the music industry, MTN Nigeria's callertunez commands a lot of revenue for both artists and the company, with reduced involvement of traditional distribution channels. The founder of Technology Times, Shina Badaru, explained that if each customer pays 50 Naira to access the callertunez and there is a customer base of over fifty million it gives an excellent idea of the level of influence it has on the economy (Akinola, 2014).

This begs the question of why the NCC would want to cripple a company that feeds a considerable amount into the Nigerian economy, outside of oil revenue.

On the surface, the fine was an attack on South Africa and South African interests. It is very easy to jump to this conclusion because of the fine also issued to Stanbic (Maake, 2015). The Nigerian government was in desperate need of foreign reserves. A number of draconian measures were instituted to preserve what is left of its foreign reserves (Financial Watch Nigeria, 2016). The big players, such as MTN came into the firing line. The impression created was that the Buhari administration had instructed all government bodies to increase foreign revenue by any means.

In the weeks before the announcement of the MTN fine, there was an article in one of South Africa's leading newspapers, the *Mail & Guardian*, essentially accusing MTN of aggressive tax avoidance (McKune and Turner, 2015). This article led to the parliamentary and political inquiry. The company has denied all allegations but has worked aggressively to improve its external stakeholder engagement. In its 2015 Integrated Annual Report, MTN notes that:

> lack of an effective strategy to engage stakeholders on their material concerns, resulting in the gradual loss of trust and confidence in MTN. In particular, key stakeholders such as governments, investors and staff feel they are not sufficiently engaged and informed about MTN's governance and operational challenges. (MTN Group Limited, 2015: 26)

Its solution to these identified challenges is to increase structured engagements and introduce 'issue management councils' across operations within the group (Ibid.).

The South African government publicly stood back from dealing with the MTN matter. But it must have been worrying, as the failure of MTN in Nigeria did not bode well for MTN's survival. Symbolically, it is apparent to those in the upper echelons of power what the failure of MTN means for South Africa. South Africa's pan-African vision would be blindsided. The most crucial move South Africa could make was repairing and improving its relationship with Nigeria. As was seen with the progress made during the Obasanjo-Mbeki era, much headway was made when both countries worked together to adopt NEPAD.

In December 2015, the NCC reduced the MTN Nigeria fine by 25 per cent to US$3.9 billion. MTN, in response, made a forward reasonable faith payment of US$250 million. However, Nigeria's lower house of parliament entered into the matter and decided the original fine should be paid (Gilbert, 2016). By the 2019 deadline, MTN had paid the balance due on the fine (NCC, 2019).

Conclusion

MTN has undoubtedly been a post-apartheid South African success story. In slightly over twenty one years, it has become a force to be reckoned with on the African continent and is fast conquering the Middle East and South East Asia. The company has gone into places considered risky, such as Sudan, Liberia and Afghanistan, and made a success of these forays.

The success of MTN and the foreign policy success of South Africa appear to be a bifurcated tree, attributing their success to the dearth of infrastructure and the third-world leadership vacuum caused by the end of the Cold War and the end of apartheid.

Throughout its various foreign policy instruments and pronouncements, the government has clarified that it looks to Africa and the rest of the Global South as its partners. The South African government helps companies expand outward by removing double taxation and providing funding through the DBSA and the IDC. The drive to move outward is further bolstered by the regimes the government creates and maintains. One that is most applicable here is NEPAD. The capstone of NEPAD is African ownership and increased economic integration on the continent. Lesetja Kganyago (2008) pointed out that companies such as MTN and Vodacom increased African integration through their operations.

This contrasts with the perception that 'business does not play a big enough role in integration' (Gruzd, 2016). White argues that proper integration has to go beyond mere treaties and agreements towards 'real' connectivity. There must be seamless movement of goods, services and people (White, 2016). It may be contended that mobile telephony has allowed Africans to move closer to their socio-economic integration goals.

It can be further argued that the operations and success of companies such as MTN entrench the perception of South Africa as a leader in Africa. This line of reasoning is that MTN is a model of South African success and dominance. Simply put, the presence and growth of MTN are part of a larger scheme to accept South Africa's apparent leadership through cultural and symbolic amelioration. South African companies, of which MTN is no exception, travel to other locales and export their products and services and South African corporate culture, particularly corporate governance standards.

7

From apartheid to democracy

The evolution of telecommunications in any country, particularly South Africa, cannot be viewed outside the interplay of international and domestic transitions. The shifts are parts of an interlocking gear that move the sector forward. The emergence of the telephone and the computer in the early 1940s marked the beginning of a new era of sophisticated information and data processing. Existing technologies such as telephones, television and radio were integrated, and new military and economic applications were developed. At the same time, two major societal shifts were occurring: the end of colonialism and the onset of the Cold War. South Africa also experienced these shifts parallel to shifts in their society: namely, the formalization of apartheid and the entrenching of a prototype for business in the telecommunications sector.

In the aftermath of the Second World War and at the dawn of the Cold War, the role of telecommunications changed globally. From being mainly a private enterprise in Europe and the United States, with state support, telecommunications became increasingly a state-owned endeavour. For example, Cable and Wireless plc was nationalized in 1947. This raised several issues, such as how to maintain technical telecommunications cooperation in a hostile political environment, address the asymmetry of telecommunications infrastructure between developed and developing nations, and ensure a more equitable representation for developing nations in information and communication governance.

These concerns were palpable for South Africa, which the government at the time had positioned as a supposed 'white' or 'European' outpost in Africa. The overall framework in which they modelled their telecommunications system was changing, and the move towards the liberalization of the sector was happening; at the same time, the country was going through internal turmoil, and there was external pressure against apartheid. South Africa's relationship with its private companies and parastatals was also changing.

Additional historical factors contributing to the success of South African MNOs in Africa

In previous chapters, I discuss how South African capital was able to expand into the rest of the continent; I also illustrate how MTN and Vodacom were able to leverage the goodwill of South Africa's successful democratic transition to internationalize. However, the seeds of the future success of South African MNOs African internalization was set in the 1960s. As African nations began to gain independence, telecommunications systems were often not considered a priority. As a result of mismanagement and insufficient investment in infrastructure, many of the telecommunications systems inherited from the colonial period fell into severe decline. For example, in the Congo, even after independence, there was little appreciation or development of its telecommunications systems, further undermined by political instability (Ganywa and Tshilombo, 1999: 122–3). Moreover, some nations remained tethered to their former colonial masters, and external influences came into play. In the Ivory Coast, for instance, the new government essentially replicated its colonial setup. From 1895 to 1960, Ivory Coast was part of French West Africa (Afrique Occidentale Française), which consisted of eight French colonial territories with Dakar as the capital. The Ivory Coast's telecommunications system was coordinated by French West Africa, and the country had no real control over its infrastructure and was not even allowed to build additional infrastructure. After independence, the government handed its telecom services over to the French-controlled company France Câble et Radio; its service expansion focused on international rather than intra-African connections (Kone, 1999: 145).

However, some African nations recognized the critical importance of improving their telecommunications infrastructure. Nigeria, for example, included telecommunications development in each of its five-year NDP. In the 1960s, it focused on expanding its telecommunications network, but this effort was later derailed by underfunding and the Nigerian civil war; nevertheless, it managed to install 26,000 telephone lines, achieving over 40 per cent of its target set in the first NDP (Ajayi, Salawu and Raji, 1999: 164).

Establishing the Pan-African Telecommunications Network (Panaftel) in the 1970s was a significant step towards improving infrastructure and connections between African nations and laid the groundwork for regional integration. However, the project ultimately failed for several reasons, including communication differences between Anglophone and Francophone countries, lack of technical skills and funding from external donors drying up; some of these donors were the ITU, UNDP and UNESCO (Pick and Sarkar, 277). Despite its eventual closure, Panaftel provided 39,000 km of radio relay links, 8,000 km of submarine cables, and 43 international switching stations. The development of the Regional African Satellite Communications Organisation (RASCOM) was part of the Panaftel project and resulted in forty-two out of forty-five member countries having international satellite stations (Ibid.).

Yet, Panaftel was doomed from the start because it relied on external donors. From the early 1980s onwards, organizations such as UNDP already sought ways to reduce funding for highly technical projects. Their rationale was that there were other sources of funding available. Lack of funds almost derailed RASCOM, a separate project conceptualized under Panaftel (UNDP General Council, 1982). Fortunately for RASCOM, it was rescued by Libyan Colonel Gaddafi, who paid 75 per cent of the funds needed to build Africa's first satellite; the remaining funds came from the African Development Bank and the West African Development Bank (Pougala, 2012).

The funding problems with Panaftel and RASCOM were part of a larger drive to push developing countries to accept funding from private capital, particularly in the telecommunications industry. Jill Hills observes that developed country governments and multilateral agencies backed this drive for private capital to invest in infrastructure and training in developing countries. In return, they expected African countries to restructure or liberalize their telecommunications sectors (Hills, 1990: 71–87).

In some countries, the situation was further exacerbated by leasing telecommunications networks to foreign companies, whose primary objective was profit maximization. This led to the redirection of international traffic to the foreign companies' countries of origin (Masmoudi, 1978: 9). Before the first RASCOM satellite was built, Africa paid Europe $500 million per annum for using its satellites for telecommunication purposes, leading to Africa having the highest telecommunications costs (Pougala, 2012: 170).

Tunisian information minister, Mustapha Masmoudi, in his working paper for the Study of International Commission on the Study of Communication Problems, commissioned by UNESCO, had the following to say explicitly about telecommunications:

> 27) The present structures and patterns of telecommunications networks between developing countries are based solely on the criteria of profitability and volume of traffic, and so constitute a serious handicap to the development of information and communication. This handicap affects both the infrastructure and tariff system.
>
> 28) With regard to the infrastructure in addition to the absence of direct links between developing countries, a concentration of communication networks is to be observed in the developed country. The planning of the infrastructure devised by former colonial powers precludes, all possibility of transmitting information beyond their frontiers (earth stations allowing only reception of televised programmes produced in industrialised countries, with no possibility of broadcasting towards these countries).
>
> 29) With regard to tariffs the situation is even more striking and in certain respect quite irrational. Designed so as to disadvantage small outputs the present tariff system perpetuates the stranglehold of the rich countries on the information flow. It is strange to say the least, that over the same distance, communications should cost more between two points within developing countries than between two others situation in the developed countries (Masmoudi, 1978:8).

During the 1980s and 1990s, private telecommunications companies began to flourish. As Hills (1994: 176) observes, formerly nationalized companies became internationalized, including state-owned enterprises with international outlooks guiding their corporate strategy. She cites the examples of BT in the UK and AT&T in the United States.

This development had significant implications for the political economy of telecommunications in Western countries. However, its impact was even more profound in Africa, where secondary economic colonization occurred as multinational telecommunications companies became increasingly politically valuable. These large companies sought to control their subsidiaries or joint ventures from their headquarters in the industrialized West. For developing countries, the forced liberalization of their telecommunications sector was challenging to accept, as it raised concerns about the potential loss of sovereignty.

Liberalization of the regulatory environment also had implications for the burgeoning MNOs. The telecommunications regulatory environment is a complex interplay of political, technical and economic settlements, which lends to crony capitalism. As Ewan Sutherland (2014a: 8) notes, the relationship between states, regulators and operators, tends to be relatively opaque in the telecommunications sectors. Initially, global MNOs viewed Africa as a high-risk market and cautiously approached it. Several players shared the risk to mitigate the political risk in any one country, as seen in the horizontal approach to internationalization. In other words, as we see with Vodafone's approach, they acquire the mobile license and ensure that they partner with an entrenched local operator. Later, single operators spread the risk across a substantial geographical footprint, growing several regional and global operator groups with a significant presence in Africa (Ibid.: 19).

South Africa's international relations and MNO internationalization

The liberalization of the telecommunications sector, alongside the neglect and inadvertent death of the fixed line services on the African continent since the end of the cold war, can be attributed to the explosion of mobile telephony services. There were insufficient fixed-line services across the continent to ensure that most people had access to telephony services for many reasons. As previously discussed, existing services became casualties of the independence movements and the Cold War. In the aftermath was a need to fill in the void. But, it is this need that further politicized the sector. Like South Africa, after decades of exploitation, African countries have been rightfully wary of *foreign* companies.

This has created an opportunity for South African mobile telephony MNCs to fill the void, with the added advantage of being perceived as '*African*'. But, the questions that continually resurface are what is an African or South African company? And does having your flag carrier company in other territories affect your ability to project your leadership capacity and soft power?

South Africa positions itself as a leader in the renewal of southern Africa and the African continent by promoting African Renaissance ideology and programs such as NEPAD. As a result, South African companies are seen as key players in driving economic integration and development on the continent.

The history of South Africa and its ability to project its power is closely linked to the embedded nature of significant capital. As demonstrated in the previous chapters, mobile telephony companies were intended to support the State's internal developmental agenda. Understandably, the nature of the agenda changed between the NP and ANC governments.

We also cannot ignore the complexity of the relationship between corporations and the South African government. While the State has implicitly supported the private sector (including MNOs) to expand their operations on the continent, there is no concrete framework to regulate their behaviour beyond South Africa's borders. This presents a dilemma for South Africa. At the same time, the State relies on the private sector to realize its vision for the continent even though it has limited control over the actions of these companies.

The Vodacom, and MTN cases are interesting for several reasons; firstly, they have both internationalized beyond South Africa's borders and present themselves as African, despite the horizontal internationalization efforts, including mergers and partnerships with external companies. Also, the shareholding and boards of both MTN and Vodacom underscore specific trends; for instance, the former seems to be home to several politically connected elites, and this model is similar to its operations on the rest of the continent, whilst the latter, due to its majority shareholding by Vodafone, seems to have taken a different path to link with political elites, which is more expected in its mergers and partnerships rather than from being at Vodacom itself. That is probably because of the considerable shareholding for a long time.

In the South African context, although some MNOs have close ties with the State, their potential to dominate the decision-making of the State is limited by the prevailing political ideology in the country. MNOs such as MTN have a privileged relationship with the government and key political players, which shapes the country's foreign relations to a certain extent. However, this does not mean that MNOs directly dictate policy to the State.

The entry of new players in the sector, such as Telkom Mobile, has also played a critical role. Initially, Telkom Mobile was expected to serve the same function as MTN in promoting the State's agenda. However, this has not been the case. Telkom Mobile does not appear to have a clear expansion strategy and has not received sufficient funding to be a competitor to MTN.

Economic diplomacy and the need for commercial diplomats

The concept of an MNC serving as a diplomat is not novel; the origins of MNCs and governments are deeply intertwined. Moreover, while economic and commercial diplomacy are often considered interchangeable, it's important to note that despite their connection, they are distinct concepts. Procassini defines commercial diplomacy as:

> the application of the tools of diplomacy to the removal of barriers to trade and investment, and to the resolution of policy conflicts arising from the globalization of the world economy. (Procassini, 2000, Online)

As defined by Van Bergeijk and Moons, economic diplomacy is centred on promoting cross-border economic activities such as imports, exports, and foreign direct investment (FDI) through open markets. It also involves using bilateral relationships to aid domestic companies facing challenges abroad, enhancing the operation of international markets, and bolstering economic security through promoting and compliance with international rules and agreements (Van Bergeijk and Moons, 2011). Economic diplomacy is a diplomatic practice that encourages economic activities within a state and leverages the relationship between the state and businesses. However, it's important to note that the identity of the dominant partner in this state-business relationship, especially in South Africa, is a subject of debate. (See Bond, 2004; Bond and Kapuya, 2006)

Therefore, commercial diplomacy is the practice of trade and investment promotion using diplomatic tools. Logically, the emphasis would be on the role of governments, even though private actors would have a role to play.

Economic diplomacy, on the other hand, encompasses commercial diplomacy. Vickers and Ajulu define economic diplomacy in the South African context as being:

> policies and activities that promote trade, FDI, tourism, and technology transfers to South Africa, and positively position the country in the world through imaging, branding, marketing and public diplomacy (domestic and international). (Vickers and Ajulu, 2008: 5)

In economic and commercial diplomacy, there are several commonalities between MNCs and states that can foster collaboration. Two primary similarities are that governments and companies are accountable to multiple stakeholders. For governments, these stakeholders include their citizens or electorate and regional and international communities. Companies, on the other hand, are responsible to their shareholders, customers, and relevant regulatory bodies. Both entities engage in public relations activities. For governments, this is known as public diplomacy.

A local firm's operations abroad can have a significant impact on their home state's foreign relations, either positively or negatively (Miller, 2009: 286). Nevertheless, South Africa arguably does not have the 'commercial diplomacy apparatus' needed to fully engage in high-level economic diplomacy (Soko and Belchin, 2009: 42).

The evolution of South Africa's economic diplomacy apparatus in the period following apartheid has not been seamless. For instance, under the Mbeki administration, the focus was not necessarily centred on economic diplomacy but rather on peace-making on the continent, including advocating for development assistance for African countries from the G7 and G8 (Fabricius, 2021: 7). However, in 2008, DIRCO, alongside the Department of Trade and Industry, developed an Economic Diplomacy Framework which was meant to provide the tools and framework for economic diplomacy. However, it was only during the Zuma administration that economic diplomacy started to take on a more prominent role, even though its exact nature remained somewhat ambiguous. Referring to the president's foreword to the Towards a 25 Review: 1994–2029, an assessment by the presidency's monitoring division, Fabricius observes that the foreign policy priorities of the Ramaphosa administration seem to be a blend of the approaches taken by the Mbeki and Zuma administrations (Ibid.: 8).

Clearly, South Africa still does not possess a robust economic diplomacy apparatus. Commentators such as Brendan Vickers and Richard Cawood (2017) have advocated for a more targeted approach to economic diplomacy, suggesting a 'formal structure chaired by the country's deputy president, with the objectives of coordinating, aligning and signalling the public and private strategies of the major South African players in the rest of Africa' (Vickers and Cawood, 2017: 6). South Africa's lack of a strong economic diplomacy machinery does not detract from the diplomatic efforts that it has made in paving the way for even more South African companies to expand on the continent than was possible during apartheid. This theme is further explored in Chapters 3 and 4. South Africa's future success relies on stabilizing its domestic political environment and harnessing its benefits from South African multinationals' prolific expansion and relative success.

Conclusion

South Africa has a long history of large corporations closely intertwined with the State's policy machinery. In the pre-1994 era, South African companies' expansion to the rest of the continent was driven by labour needs, especially by the agricultural and mining sectors, and a desire to conquer new markets. Business interests often overlapped with those of the state, as was the case of the South African state involvement in the Lilongwe Capital Project in the 1960s.

As Firoz Khan and Seeraj Mohamed remind us, from 1974 to 1994, leaders of white businesses worked very closely with top military officers of the NP and the leaders of white-owned businesses. Their common goal was to resist what they identified as the 'Total Onslaught' from communists (Khan and Mohamed, 2023: 724). What Khan and Mohammed discovered in the archival records indicates a reciprocal relationship between business and the State. Business leaders wanted access to larger markets and cheap labour; they also wanted to stem the tide of socialism and communism that appeared to be sweeping across the continent, particularly in the 1970s. The State apparatus wanted the continuation of the existing racial and political hierarchy. In the transition to democracy, many white business leaders got involved in peace negotiations by initiating a series of meetings with the apartheid government and liberation movements, but necessarily for altruistic reasons. In the post-1994 era, the State and significant capital relationship has been complex and fraught with ideological and racial tensions. The State-business-labour nexus further complicates this relationship, as these three actors often appear to conflict.

The SBR relationship can be symbiotic, underpinned by shared interests converging on several issues. There is an additional incentive for closer SBR in telecommunications, in particular, because of the economic, social and developmental implications. Throughout this book, we observe how the evolving dynamics between South Africa, MTN and Vodacom have shaped policies like BEE and influenced South Africa's reputation. The role of big business, particularly those that operate beyond South Africa's borders, in expanding the country's political reach and increasing its imprint is significant. South Africa's peaceful transition to democracy in 1994 enhanced its image abroad and on the continent. In the period 1993 to 1998 in particular, it became a symbol of political stability and an example for other countries. However, the narrative of mobile telephony in Africa and globally is far more complex and multifaceted.

For instance, in examining mobile telephony in China, Jia Lu and Ian Weber's research shows how the state uses nuanced forms of influence to create a 'relational contract' with partners in the Chinese telecommunications and information industry and its citizens (Lu, Jia and Weber, 2007). They essentially argue that this contract serves as a basis for executing a controlled commodification strategy; it helps redefine and set clear limits within the private and public sectors, aiming to balance and manage the often conflicting economic, social and political goals. However, 'relational contracts' are not just the domain of authoritarian states; in the relations between Spain and Telefonica, Angela Calvo illustrates how the relationship helped the company move from one of the weakest operators in Europe in the 1980s to one of the most efficient by 2009. State support improved Telefonica's market positioning, allowing for successful internationalization, especially in Latin America, and organizational restructuring (Calvo, 2014: 6–20).

Throughout the 1980s and 1990s, the telecommunications industry has undergone massive reforms with implications for SBR. As illustrated in previous chapters, these reforms have implications for South African MNOs, including MTN and Vodacom. As mentioned earlier, accountability is one domain with notable convergence between States and MNCs, particularly MNOs. States are accountable to their citizenry or electorate and regional and global communities. Similarly, corporations are obligated to their shareholders, clientele and pertinent regulatory institutions. In a developmental context, as we see in the South African case, there is a further responsibility to provide access to services to poorly resourced communities. But, it is the process of ensuring accountability that can lead to a disconnect between the regulatory impulse of different sectors of the State and the ability of an MNO to operate, efficiently. Moreover, the struggle to harmonize regulatory frameworks with political agendas is a common issue encountered by all nations. For instance, US president Bill Clinton and his vice-president, Al Gore, signed what was then a ground-breaking act, the Telecommunications Act of 1996, which was meant to loosen regulatory strictures to foster greater competition in the sector and ensure universal coverage. Gore, in particular, was an advocate for increased public access to communication tools, specifically the internet since the 1980s. However, the United States, operating under a federal system, has a regulatory framework that is more complex compared to Europe. As Robert G. Harris and C. Jeffry Kraft point out, the telecommunications market in the United Sates, despite its liberal market ideals, has been consistently shackled by the dual state and federal structure. This structure allows for regulation by the US Congress, the Federal Communications Commission (FCC) and state regulators, with additional powers granted under the Telecommunications Act (Harris and Kraft, 1997: 96, 109; Cave, Genakos and Valletti, 2019). Although the act represented a major overhaul of the industry in over sixty-two years and was the first time that the internet was included in broadcasting, its aim to allow for more competition, including in mobile telephony, was not achieved (McCabe, 2016).

State would acquiesce to the desires of MNCs because of the need for access to capital. But, in reality, the relationship is a little more complicated, but can be mutually beneficial. Pepper D. Culpepper skilfully addresses the conceptual conflicts related to structural power that both the State and MNCs possess, pointing to the interconnected relationships between the State and MNCs. He points out that implementing policies that align with business interests is not necessarily a result of governments yielding to structural power. Instead, it could be because such policy changes satisfy the preferences of their constituents (Culpepper, 2015: 395). Put differently, businesses gain from policies that are designed to meet the needs and/or national interests of a State's citizenry.

In South Africa's case, owing to how the country has positioned itself with its African agenda, its stakeholder include its citizens and those of its immediate neighbourhood and the continent at large. Of course, this comes with tensions with South African MNCs, who sometimes expect a different level of engagement. The tensions are intensified by the public diplomacy role that MNCs unwittingly have to play in response to the profits they gain from the State's policies, including its image and perception in the global arena.

In 2002, approximately at the halfway point of Mbeki's term in office, Anglo American's CEO, A. J. 'Tony' Trahar, expressed frustration with the contentious state-business relationship in South Africa. He noted that the government's definition of partnership was one where it proposed and disposed of issues, pausing only to solicit funds from the private sector (Trahar, 2002: 6). However, many companies have been able to access previously impenetrable markets on the continent by piggybacking on South Africa's foreign policy engagements. In the case of mobile telephony, British Vodafone bought a controlling interest in Vodacom but kept its South African brand identity due to its strong recognition and ability to tap into existing resources and local knowledge (Luiz and Stephan, 2012: 626).

Moreover, having headquarters in Gauteng[1] brings additional benefits for companies, as they are closer to the headquarters and regional offices of international bodies and the seat of government. Programs such as NEPAD, and increasing African integration (including the African Continental Free Trade Area, which became effective in 2019), create more opportunities for South African-based firms. Most of South Africa's ICT firms are based in Gauteng (Diop, 1996), and foreign firms based in South Africa may also accrue similar benefits.

A key feature of internationalization, or Africa-focused expansion, has been using the mantle of the African Renaissance. The idea of a season of 'African Renewal' or a 'Renaissance' is not a new one; the most famous articulation is the

1 Gauteng is South Africa's economic hub and financially, the most influential province.

one by Senegalese scholar Chiek Anta Diop (1996) in his collection of essays, *Towards the African Renaissance: Essays in African Culture & Development, 1946–1960*. We see an idea move towards the crystalizing vision of policy when Thabo Mbeki, as deputy president, presented his 'I am an African' Speech in 1996 (Mbeki, 1996), which echoed Pixley ka Isaka Seme's speech on the 'regeneration' of Africa in 1906 (Seme, 1906). The contours of Mbeki's future Africa-centred policy were detailed in an African Renaissance statement in 1998 (Mbeki, 1998). These statements and other intellectual work around the African Renaissance project, including the ANC's African Renaissance conference in 1998, served as the intellectual backdrop to South Africa's foreign and economic policy towards the rest of the continent (Netshitenzhe, 1998; Makgoba, 2020).

However, as James Hamill argues, the 'rhetoric on the African Renaissance' laid the groundwork for its failure by fostering inflated anticipations about its potential influence in Africa (Hamill, 2016: 91). To some extent, his observation is correct. South Africa had unwittingly painted itself into a corner by positioning itself (or being pushed into the position by its allies and interested Western parties) as a leader and a moral beacon in Africa. That positioning assumes a conducive domestic environment and a financial and political appetite to sustain such a position. Even though there has been a move towards matching rhetoric to action, South Africa has still to bring its vision to fruition. However, this has not prevented South African companies and other corporate entities from leveraging the momentum and enthusiasm generated by South Africa through the African Renaissance movement.

Businessman Tony Elumelu suggests that 'Africa's renaissance lies in the confluence of the right business and political action' (The Tony Elumelu Foundation, n.d.). However, this ideology has been challenged by those who reject globalization and capitalism as the most appropriate ideology for the global political economy, particularly Africa, as it puts power in the hands of an elite group or business class.

Yet, this ideology underscores why MTN and Vodacom cling to their identities as African companies with South African DNA. The genesis, and future internationalization, of these rival MNOs cannot be viewed outside of the prism of BEE, B-BBEE and South Africa's leadership on the continent. Due to JSE listing mandates, Vodacom and MTN have been forced to use local suppliers and build up local talent. On the latter, MTN has been more successful than Vodacom. This modus operandi is replicated – to varying degrees across its various operations.

The African internationalization of both companies demonstrates how their expansion and pervasiveness are facilitating the integration of the African economy. MTN has been the most explicit in exporting South African ideals and corporate culture and adapting them to their host countries. As shown in the chapter on the MTN group, every MTN location has followed a similar model with critical South African ideals as pillars. One of these pillars is the indigenization of the local entity of the MNC through shareholding and even by listing shares on the local stock exchange. South African corporations, particularly in the mobile telephony sector, have been instrumental in building and enhancing an ecosystem to support their operations.

This ecosystem ranges from the grassroots level, with street vendors selling recharge vouchers, to higher levels, with local engineers and other professionals whose services are utilized directly or indirectly by these companies. A prime example is MTN Rwanda, where the economic and IT ecosystem was enriched when negotiations for a GSM license began for MTN to operate in the country. New IT businesses learned how to negotiate with large multinationals, and technology transfers enabled other companies to emerge and strengthen the telecommunications infrastructure in Rwanda.

A three-way relationship exists between the host state, the home (or flag) state and the MNC. In the period following 1994, South Africa, the home state in this context, has not acted as much of an interlocutor as it did during the apartheid era. As highlighted in Chapters 5 and 6, particularly concerning taxation issues, there are crucial aspects that need attention. This is especially true as we transition into a digital economy in the twenty-first century.

The tax systems in many African nations need to adapt to the rapidly changing digital economy (Ndajiwo, 2020: 8–13; Mpofu, 2022). Approximately US$500 billion in tax revenue is lost annually by developing countries due to profit-shifting strategies employed by digital platforms (Cobham and Janský, 2018: 221; Onuoha and Gillwald, 2022: 5). However, MNOs, MTN and Vodacom bear the brunt of the pressure regarding the tax burden in this telecommunications sector. Perhaps it is due to their size and visibility – however, the impasse over taxes between the Nigerian government and MTN in 2020 and 2023. MTN has had a similar issue with Ghana unsuccessfully attempting to claim $773 million back taxes, and Vodacom dealt with a similar problem in the Democratic Republic of Congo (Bloomberg, 2023).

MNOs are still MNCs, and as much as they are development instruments, their capital significantly contributes to both the home and host states through FDI. This investment not only boosted the GDP of the host state but also resulted in substantial corporate tax revenue for the home state. Specifically, MTN Group paid R8.1 billion in taxes, and across its operations, the total reached R31.5 billion in 2022 (Brederode, 2023). Despite these substantial contributions, MNOs need to rationalize their expenditure. Mike Rogers and Xavier Pedros highlight that the mobile industry necessitates considerable upfront investment in spectrum licenses, equipment acquisition, network development and points of sale. This is especially pronounced in sub-Saharan Africa, a region with a predominantly rural demographic. The financial burden associated with the extension and enhancement of mobile networks in this area is substantial. Approximately half the population, equating to nearly 500 million individuals, lives beyond the coverage of a 3G mobile network. This deficit highlights the economic difficulties associated with deploying infrastructure in remote areas. In these regions, costs can double while the revenue opportunities may only be a tenth of what can be achieved in urban areas (Rogers and Pedros, 2017: 28).

Moreover, the benefits of the internationalization of South African MNCs go beyond mere firm-level benefits such as increased markets and state-level benefits such as increasing 'soft power', taxation income and bolstering the external perception of its leadership. As illustrated in Chapter 3, South Africa's SBR has always been challenging. Yet, we cannot forget that the apartheid State did assist the business community in expanding beyond its borders. The post-1994, ANC-led government also stepped in to assist private and state-owned corporations to grow their imprint on the continent. Their reasoning was partly ideological, including fostering regional integration and shifting the negative perception of South Africa on the continent. Vickers and Cawood remind us that the government of South Africa consciously created an incentive scheme to boost and advance foreign direct investment flowing from South Africa to other African nations. In 2014, this outward investment was valued at R200 billion, making up 32 per cent of all such investments, predominantly in southern Africa (Vickers and Cawood, 2017: 353–4).

We also cannot ignore that internationalization benefits the state's diplomatic endeavours. As Fritz Nganje and Letlhogonolo Letshele note, the international expansion of South African companies could greatly benefit the country's foreign policy, especially as optimizing economic diplomacy becomes more salient in a shifting global economy (Nganje and Letshele, 2020: 246).

As I illustrate throughout this book, the story of the political economy of mobile telephony – precisely that of MTN and Vodacom – is one that is intertwined with the history and political transformation of South Africa. Mobile telephony has played a significant role in shaping the business landscape of South Africa. It has helped forge a Black empowerment business model that linked the state, capital and the bourgeoning Black elite. This state-business nexus has profoundly impacted South Africa's foreign policy in Africa, boosting its soft power projection.

However, these new private firms must negotiate their profit motives with the state's expectations and incentives. While South African MNCs, as private actors, contribute to consolidating South Africa's leadership role and aligning its economic interests with its foreign policy goals, it is up to the state to begin explicitly driving the agenda and setting the terms of engagement on the rest of the continent.

Appendix A

Memorandum: South Africa, a little more history

Much of the historical background is already covered. However, it is essential to paint a broader picture of South Africa's involvement in telecommunications via mobile telephony, which was more a continuation of the trailblazer role it had taken regarding telecommunications. This memorandum illustrates that South Africa's influence has always been inextricably linked to the rest of the continent. Similarly, the growing political voice of other African countries is linked to South Africa's international prowess.

In 1998, Keohane and Nye recognized that information technology had reshaped international relations. This assertion still holds, and nowhere is this more evident than within the International Telecommunications Union (ITU).

The ITU is the regulatory regime for Information and Communication Technologies (ICT) and was one of the first multilateral organizations created in 1865. It later became a specialized UN agency in 1947. Contemporary discourse on Africa's relationship with ICTs leans heavily towards concern about bridging the development gap whilst overlooking Africa's potential influence in governance structures. South Africa is one of the ITU's oldest members, joining in 1881. Although sometimes clandestine, it held the position of a regional representative within the ITU. Global backlash against the country's policy of apartheid derailed its position, but post-apartheid, the country appears to have reclaim some of its former standing.

Primarily utilizing archival material from the ITU and South Africa's national archive, I consider South Africa's relationship with the ITU in two distinct phases: pre- and post-1994. The distinction is made as 1994 is regarded as a watershed, the year South Africa became a fully inclusive democracy. The political shift in the country affected its engagement with the international community, including the ITU.

The ITU is an essential site of analysis, where national interests converge and the dimensions of the international system shift. It is the premier regime governing ICT, the cornerstone of geopolitical developments. Yet, there is a dearth of contemporary and historical discourse on Africa's, let alone South Africa's, relationship with the ITU. Most contemporary investigations fall within three camps: the construction of the ITU, which is primarily historical (Michaelis, 1965; Maclean, 2003), the international functioning and reformation of the ITU (Codding, 1952, 1991, 1994; Lyall, 1994, 2011) and the role of ITU as a regulator (Codding, 1994; Cowhey, 1990; Maclean, 2003; Fari, Balbi and Richeri, 2015).

Background

Unlike other African countries, South Africa has actively participated in all aspects of telecommunications – telegraphy, telephony and radio– for a considerable time. The reasons for this were its character as an extension of the empire and the technical know-how that was present in the country at the time. The telegraph was introduced to the African continent in the 1870s, and the British South Africa Company, founded by Cecil John Rhodes, built telegraph and railway lines primarily in Southern Africa to ease communications with the Crown. South Africa had its first telephony network in 1891. Whilst South Africa was still a Union, telegraph and telephony were administered by the South African Post and Telecommunications (SAPT) service. However, the SAPT was controlled by the British Treasury, and the Exchequer controlled revenue and spending (Horwitz, 2001: 75–6).

Moreover, when South Africa wanted to engage with the Marconi Company, it had to get permission through a British parliamentary process despite already having engaged with the Marconi Company (House of Commons Debate, 1913; House of Commons Debate, 1924; South African National Treasury, 1925).

After South Africa became a republic, it became clear that telephony infrastructure and human resource deployment were racially biased. The Republic had leaders and technocrats who appreciated the importance and potential of telecommunications. One such technocrat was Hendrik Johannes van der Bijl, the first chief executive officer of Eskom, who was an engineer and administrator with many patents, including creating an essential component for wireless telegraphy, the thermionic valve (Vermeulen, 1998: 2454). He invented a telephone that could scramble calls, which was helpful during the war when secure telephony lines were crucial (Jacobs, 1948; Vermeulen, 1998: 2451). He also created a rudimentary fax machine (Vermeulen, 1998). The administration ensured that South Africa continued to be involved in creating the telecommunications regulatory framework regionally and internationally and improving telecommunications access for its white population.

South Africa's policy of apartheid made it an international outcast, even within the ITU. As a result, South Africa was expelled from the ITU in 1965 during the Montreaux Plenipotentiary Conference. This conference was also significant because it revised the structure of the ITU. Most notably, the administrative council was expanded, and regional telecommunications unions were established. This was an important development for African countries because they could concentrate on the issues of African nations under this new arrangement.

While South Africa was isolated from the ITU and the international community at large, many significant developments occurred within the ITU and on the continent that had profound implications for the geopolitics of telecommunications.

South Africa's fall from grace

South Africa's fall from grace in the ITU closely parallels what was happening within the UN. Its policy of apartheid regularly put it in the crosshairs of the United Nations from 1946. Moreover, the issue of South West Africa (later called Namibia) further intensified the calls against South Africa. South West Africa was originally a German colony, and after the First World War, South Africa received a mandate over the territory (Dugard, 1973: 28). By 1962, South Africa was hauled in the International Court of Justice as it refused to accept that it no longer had a mandate over the territory (South West Africa Cases (*Ethiopia v. South Africa*; *Liberia v. South Africa*, 1962: 319).

By then, the ITU was entirely part of the UN system and could not remain silent. South Africa had begun representing South West Africa at the ITU and it was referred to in official documents as the Union of South Africa and Territory of South West Africa. A mere two weeks after South Africa's mandate over what was then South West Africa was revoked, it attempted to deposit an agreement not only on its behalf but also on behalf of its former territory (Ronen, 2011: 121). South Africa's provocative action caused an uproar amongst members of the ITU, particularly those from developing countries[1]. Ultimately, it led to the members agreeing that South Africa no longer had the right to represent South West Africa (ITU, 1965), precipitating its active Union membership.

By the time of the Montreaux Plenipotentiary Conference, the presence of South Africa was not welcome. Several countries registered their distaste for South Africa's presence in the convention. They stated emphatically that 'subsequent ratification of that document by their respective Governments in no way implied recognition by these States of the present Government of the Republic of South Africa, and it did not entail any obligation towards that Government' (ITU, 1965: Para V, 88). Moreover, Resolution 44 and Resolution 45 of the convention effectively barred South Africa from active participation in the Union. The previous resolution barred South Africa from participation in regional conferences for Africa and the latter barred the country from the Montreaux Plenipotentiary Conference.

South Africa and the ITU post-1994

As illustrated, the political economy of telecommunications thrived despite South Africa's absence. However, for various reasons, the political gains of

[1] The countries were Algeria (Algerian Democratic and Popular Republic), the Federal Republic of Cameroon, the Central African Republic, the Democratic Republic of the Congo, the Republic of the Congo (Brazzaville), the Republic of the Ivory Coast, the Republic of Dahomey, Ethiopia, the Gabon Republic, Ghana, the Republic of Guinea, the Republic of Upper Volta, Kenya, the Republic of Liberia, Malawi, the Malagasy Republic, the Republic of Mali, the Kingdom of Morocco, the Islamic Republic of Mauritania, the Republic of the Niger, the Federal Republic of Nigeria, Uganda, the United Arab Republic, the Somali Republic, the Republic of Rwanda, the Republic of Senegal, Sierra Leone, the Republic of the Sudan, the United Republic of Tanzania, the Republic of Chad, the Togolese Republic, Tunisia and the Republic of Zambia.

developing countries, particularly African countries, did not necessarily translate into closing the stark telecommunications access gap. Upon the ending of apartheid and the ushering in of a fully democratic constitution, South Africa was warmly welcomed back into the international arena. At the 1994 Kyoto Plenipotentiary Conference, the ITU reaffirmed its commitment to regional divisions (ITU, 1994: Res. 25) and welcomed South Africa back into active participation in the ITU. This move was significant as South Africa had been outside international telecommunications governance for so long that it could now play a more meaningful role. Concerning closing the infrastructure gap or bridging the digital divide, the forum for that to take place was ITU. At the 1998 Minneapolis Plenipotentiary Conference, the proposals for hosting the World Summit on the Information Society (WSIS) ultimately led to WSIS becoming a United Nations General Assembly mandate with the ITU as the manager.

Although South Africa was absent for many years from the formal information and communications ecosystem, the country's leadership potential was still recognized. Soon after South Africa's democratic dispensation took over in 1994, the country was invited to address the G7 Ministerial Conference on the Global Information Society; it was the only African country invited. Throughout the discussions at the conference and within Mbeki's keynote address at the opening of the Brussels conference, he stressed the need for the information boom to be universally beneficial. Interestingly, he boldly proposed that South Africa take the lead by hosting an information society conference for the developing world. Drawing on the lessons that South Africa had learned in the lead-up to its first elections as well as new lessons learned about governance, Mbeki was unquestionable when stating:

> We believe that South Africa can act as one of the bridging societies concerning the realisation of the common objective of bringing together, in a mutually beneficial way, the interests, assets and aspirations of the developing world and the technological and financial capacities that reside in the developed world. (Mbeki, 1995)

Building upon the groundwork laid by other developing nations, it was evident that the country was resuming the mantle of leadership it assumed in the early days of telegraphy and international telecommunications governance.

Echoing Mbeki, President Nelson Mandela underscored the importance of inclusivity for the development of international telecommunications policy and asserted that,

> eliminating the distinction between information rich and information poor countries is also critical to eliminating economic and other inequalities and to improving the quality of life of all humanity. (Mandela, 1995)

Mandela's comments were made at the very first event that South Africa attended when its membership with ITU was restored, the World Telecommunications Conference and Exhibition in Geneva. Unfortunately, even when South Africa hosted the same conference in 2018, the inequalities to which Mandela and Mbeki called attention still existed. Two major ITU conferences took place in 2018: the Telecom World Exhibition in Durban and the Plenipotentiary Conference in Dubai. Both events promised to change the geopolitical telecommunications environment significantly and was an opportunity for South Africa to underscore its leadership abilities and technical capabilities.

South Africa secured the right to host the telecoms conference because no African country has been chosen to host the event since it first began in 1971. South African Telecommunications and Postal Service minister framed South Africa's hosting of the event as an African event. He was reported as saying:

> Now is the time for Africans to host the global event to enable African countries to fully benefit from the fourth industrial revolution, which is knowledge-driven. ITU in Africa will further boost technology development in Africa and empower technology startups and small businesses to catch up with global trends where technology is driving global advancements. (ThisDay LIVE, 2017)

Reforms in the ITU

Other than the expulsion of South Africa at Montreaux in 1965, several reforms took place dating back to the Atlanta meeting. The reforms between 1965 and 1989 were significant and had wide-ranging implications for newly independent African countries.

Once it was agreed that the ITU would become a part of the UN in 1947, two years after the creation of the UN, the ITU signed an agreement to become a specialized agency of the UN. The deal was signed in the United States at Lake Success (E/RES/90(V)). It came into force only on 1 January 1949. The Atlantic City agreed that it would establish an administrative council, and at the Buenos Aires conference, the actual tasks of the officials at the council were defined.

Codding (1989: 14) observes that developing nations have benefited from the open nature of the ITU for several reasons, the most important being that from 1965, with the growing number of newly independent states, the membership of the ITU ballooned to 124. This allowed for a collective voice to emerge in the call to increase the assistance the UN and the ITU provided.

By 1973, the Malaga-Torremolinos Plenipotentiary Conference was virtually dominated by developing countries, allowing it to steer the agenda towards increased development support (Codding, 1989: 15). In 1982, at the Nairobi Plenipotentiary Conference, the proposals for financing technical assistance that had arisen at every conference since 1965 had emerged again as was the call for the ITU to be segmented into regions. The regional focus of the ITU was the more successful of the two proposals (Codding, 1989: 16).

The 1989 Nice Plenipotentiary Conference established the Telecommunications Development Bureau (BDT)[2] and was created at the Nice Plenipotentiary Conference (ITU, 1989: Res. 17). The Bureau intended to develop policies to assist developing countries in improving their telecommunications capacity. In 1992, the ITU established its Regional Office for Africa, based in Addis Ababa, under the BDT.

The most wide-ranging reforms occurred at the additional plenipotentiary conference in Geneva in 1992. The conference adopted a new constitution (the first since 1865) and settled on convening a plenipotentiary conference every four years (ITU, 1992: article 8). For developing countries, one of the most critical structural change was the development of three bureaus: radio communications, telecommunication standardization and the telecommunication development bureau (ITU, 1993). Each bureau would have

[2] Bureau de développement des télécommunications.

an advisory group from member states and the private sector. The development bureau just backed the decision that was made in 1989.

The strength of developing nations was buoyed by the sheer numbers and a growing desire to be taken seriously and create a more equitable global world order.

Conclusion

The question of why South Africa has yet to be more assertive in its leadership of the African ICT sector can be raised in the context of the history of telecommunications on the continent. A plausible explanation is that African nations have not always been kindly disposed towards South African leadership. A clear example of this was the New Partnership for Africa's Development (NEPAD) internal contestation around ICT infrastructure.

NEPAD adopted the Protocol on Policy and Regulatory Framework for NEPAD ICT Broadband Infrastructure for Eastern and Southern Africa, commonly called the Kigali Protocol, in 2008 after it was first created in 2006. The protocol aims to facilitate the development of regulatory frameworks and a special-purpose vehicle that operates and maintains a regional broadband network. To this end, NEPAD developed Uhurunet, a submarine cable system, and Umojanet, a terrestrial network. By 2009, Baharicom Development Company (BDC), the NEPAD-led submarine company, signed a memorandum of understanding with the ACE Consortium[3] to help build the Africa Coast to Europe high-capacity submarine cable. Also, under the Kigali Protocol, regional infrastructure initiatives such as COMTEL, the Southern African Regional Information Infrastructure and the East African Backhaul System were incorporated (UNECA, 2017: 30). The Kigali Protocol promotes an open-access network with the intention that landlocked countries would be penalized by countries where the cable lands.

The staggering number of private and state-led consortiums has accelerated the development of backbone infrastructure, particularly submarine cables.

[3] This consortium comprises twenty-five investors led by France Telecom/Orange. One of the investors is the South African state-owned company, Sentech.

These relationships need to be optimized, and great opportunities are missed. The 2003 Eastern African Submarine Cable System (EASSy)[4] is one such regional project. EASSy was initially envisioned as part of the NEPAD vision for developing the continent's backbone infrastructure (NEPAD e-Africa Commission, 2005). To that end, NEPAD pursued funding for the project (Muller, 2007). However, Kenya became a stumbling block as it was not in agreement with having to increase its government's regulation of the sector, and it did not agree with the primacy South Africa had, not only within NEPAD but among the South African companies within the EASSy Consortium (Gedye, 2007). Eventually, owing to the contentious relationship, NEPAD withdrew from EASSy and put forward its project, Uhurunet.

The bickering that saw the relationship between EASSy and NEPAD end is a mere reflection of the challenges facing regional integration, projects and connections stymied by attempts at political badgering.

[4] It aims to connect the east coast of Africa to the rest of the world.

Appendix B

Key historical ICT indicators, 1965–2012

Table 5 Telecommunications Profile. South Africa, 1965–2009

	1965	1970	1975	1980	1982	1985
Telephone faults (per 100 mainlines)	..	..	..	..	..	..
Telephone lines	755,000.0	916,000.0	1,156,000.0	1,632,000.0	1,936,620.0	2,297,773.0
Telephone lines (per 100 people)	3.8	4.1	4.7	5.9	6.7	7.3
Telecommunication equipment – import (current US$)	..	..	..	..	..	..
Telephone (mainlines and mobile phone) subscribers (per 1,000 people)	38.1	41.5	46.7	59.2	66.8	73.4
Annual investment in telecommunications (current LCU)	..	77,000,000.0	167,000,000.0	378,880,000.0	784,540,992.0	1,376,299,008.0
Mobile and fixed-line telephone subscribers	755,000.0	916,000.0	1,156,000.0	1,632,000.0	1,936,620.0	2,297,773.0

Source: Adapted from World Bank (2023).

1990	1993	1994	1999	2000	2005	2009
..	..	80.0	52.2	51.8	..	..
3,315,022.0	3,593,851.0	3,775,355.0	5,492,838.0	4,961,743.0	4,729,000.0	4,319,800.0
9.4	9.6	9.9	12.8	11.3	10.0	8.8
18,700,000.0	55,500,000.0	32,900,000.0	..	165,903,568.0	142,578,128.0	..
94.3	97.0	107.5	248.8	302.3	819.7	1,029.1
1,515,000,000.0	2,382,000,000.0	3,500,000,000.0	11,900,000,000.0	12,100,000,000.0	..	14,646,000,000.0
3,320,702.0	3,633,851.0	4,115,355.0	10,680,838.0	13,300,743.0	38,688,958.0	50,755,800.0

Table 6 Africa ICT Profile (Excluding South Africa), 1994

Country Name	Annual investment in telecommunications (current LCU)	Mobile and fixed-line telephone subscribers	Telecommunication equipment – import (current US$)	Telephone (mainlines and mobile phone) subscribers (per 1,000 people)	Telephone faults (per 100 mainlines)	Telephone lines	Telephone lines (per 100 people)
Mozambique	62103470000	57490	0	3.730844218	57	57490	0.373084422
Cote d'Ivoire	26641000000	103268	0	7.24486795	65	103268	0.724486795
Senegal	19411000000	72187	0	8.864762242	39.35	72089	0.885272757
Cameroon	14185000000	59800	0	4.400391797	0	58200	0.428265556
Zambia	12500000000	79786	0	9.178612749	123.84	79786	0.917861275
Uganda	9243280000	30449	0	1.507866518	100	30449	0.150786652
Benin	8114910000	23945	0	4.374940734	90	23945	0.437494073
Burkina Faso	6557000000	26328	0	2.531288984	88.77	26328	0.253128898
Algeria	3805000000	1123757	0	40.494163	74	1122409	4.044558833
Kenya	3508203000	230512	0	8.651915237	196	228522	0.857722363
Tanzania	3498850000	88686	0	3.051380386	210	88315	0.303861555
Morocco	2476000000	1020794	0	38.48625375	63	1007000	3.79661886
Chad	2375080000	4733	0	0.69748161	97.18	4733	0.069748161
Nigeria	2314800000	381515	0	3.550541393	0	368715	0.343141913
Togo	2100000000	21400	0	5.373098952	20	21400	0.537309895
Burundi	1851610000	16166	0	2.68666425	28	15788	0.262384357
Niger	1600000000	12453	0	1.403687777	0	12453	0.140368778

Sierra Leone	1240630000	15754	0	4.023942381	10.8	15754	0.402394238
Mauritania	1166635854	8426	0	3.781098367	206	8426	0.378109837
Egypt, Arab Rep.	1003435000	2463532	0	40.36454205	0	2456161	4.02437695
Mali	773789000	15203	0	1.589568466	0	15203	0.158956847
Central African Republic	741050000	7373	0	2.270392578	0	7373	0.227039258
Djibouti	672906410	7516	0	12.25663262	0	7516	1.225663262
Mauritius	601700000	135149	0	121.4444766	104.8	129443	11.63170825
Zimbabwe	502000000	134812	0	11.75357493	190	134812	1.175357493
Cape Verde	463000000	18450	0	47.90950945	36	18450	4.790950945
Sao Tome and Principe	310080000	2460	0	19.60346806	27	2460	1.960346806
Sudan	226905235.7	64000	0	2.688683964	20	64000	0.268868396
Tunisia	123164000	476962	0	54.10554257	92	474253	5.379823944
Botswana	102607000	50447	0	32.62192401	34.8	50447	3.262192401
Namibia	69604000	69784	0	43.50989641	0	69784	4.350989641
Swaziland	54340000	18189	0	19.25976595	206	18189	1.925976595
Ethiopia	46944485	137731	0	2.491468415	0	137731	0.249146842
Gambia, The	32850000	19118	0	17.4597753	46	18306	1.671820518
Seychelles	17963959	12110	0	163.815537	45	12110	16.3815537
Eritrea	14340000	15192	0	4.768267294	0	15192	0.476826729
North Africa	0	5356561	0	41.38817695	74	5331339	4.119329583
Congo, Rep.	6362000	20553	0	7.729748758	0	20553	0.772974876
Lesotho	5600000	15712	0	8.919377051	0	15712	0.891937705
Libya	0	264000	0	56.32473428	0	264000	5.632473428
Angola	0	54844	0	4.670372717	0	53020	0.451504561

Continued

Table 6 *Continued*

Country Name	Annual investment in telecommunications (current LCU)	Mobile and fixed-line telephone subscribers	Telecommunication equipment – import (current US$)	Telephone (mainlines and mobile phone) subscribers (per 1,000 people)	Telephone faults (per 100 mainlines)	Telephone lines	Telephone lines (per 100 people)
Ghana	0	53343	0	3.222196751	176	50007	0.302068487
Madagascar	0	33975	0	2.669287217	120	33675	0.264571735
Malawi	0	33081	0	3.385227673	0	33081	0.338522767
Gabon	0	33979	0	32.18601373	0	31398	2.97412066
Congo, Dem. Rep.	0	0	0	0	0	36000	0.084407483
Somalia	0	15000	0	2.313305454	0	15000	0.231330545
Guinea	0	10120	0	1.395947174	1150	9308	0.128394035
Rwanda	0	10000	0	1.770442324	0	10000	0.177044232
Guinea-Bissau	0	6935	0	6.287380905	0	6935	0.62873809
Liberia	0	4500	0	2.205966699	0	4500	0.22059667
Comoros	0	4272	0	8.855703035	0	4272	0.885570304
Equatorial Guinea	0	2515	0	5.876099578	50	2515	0.587609958

Source: Adapted from World Bank (2023).

Table 7 Growth in the Number of Vodacom and MTN South Africa's Subscribers and Market Share, 2010–20

	Vodacom South Africa		**MTN South Africa**	
	Subscribers (million)	**Market share (%)**	**Subscribers (million)**	**Market share (%)**
2011	31.727	49.6	22.03	34.3
2012	30.581	44.7	25.421	37.2
2013	30.964	40.3	25.700	33.4
2014	31.379	39.6	27.993	35.3
2015	34.103	39.2	30.588	35.2
2016	36.375	44.7	30.764	37.8
2017	41.602	47.7	29.516	33.9
2018	43.838	47.8	31.191	34.0
2019	44.341	45.7	28.890	29.8
2020	44.312	46.7	32.890	33.7

Source: Adapted from Lancaster (2022: 583, 589).

Appendix C

Key mobile telephony statistics

Table 8 MTN Group Ltd Profile (as of 31 December 2022)

Country	Ownership (%)
Afghanistan	100
Benin	75
Botswana	53.1
Cameroon	80
Eswatini (Swaziland)	30
Ghana	84.3
Guinea-Conakry	75
Guinea-Bissau	100
Iran	49
Ivory Coast	66.8
Liberia	60
Nigeria	75.7
Rwanda	80
South Africa	100
South Sudan	100
Sudan	85
Uganda	83.1
Zambia	89.8

Source: MTN Group Ltd (2022).

Table 9 MTN Subscribers ('000), 2022

Country	1Q 21	2Q 21	3Q 21	4Q 21	1Q 22	2Q 22	3Q 22	4Q 22
South Africa	32 214	32 450	33 185	34 995	34 509	35 346	35 878	**36 538**
Postpaid	7 008	7 039	7 221	7 359	7 526	7 808	7 971	**8 277**
Prepaid	25 206	25 411	25 963	27 635	26 982	27 537	27 908	**28 261**
Nigeria	71 512	68 921	67 542	68 471	70 185	74 156	74 099	**75 635**
SEA	33 790	34 423	34 748	35 118	34 769	34 831	35 897	**36 499**
Uganda	14 667	14 932	15 295	15 716	15 920	16 255	16 701	**17 170**
Rwanda	6 162	6 475	6 428	6 435	6 468	6 585	6 792	**6 816**
Zambia	8 374	8 295	8 260	8 056	7 297	6 812	7 117	**7 112**
South Sudan	1 829	1 937	1 966	2 062	2 198	2 309	2 387	**2 484**
Botswana (joint venture)	1 764	1 784	1 792	1 815	1 831	1 831	1 834	**1 836**
Eswatini (joint venture)	995	999	1 006	1 034	1 055	1 039	1 067	**1 081**
WECA	64 912	65 525	66 179	68 112	70 202	70 423	72 033	**72 619**
Ghana	24 982	24 866	25 223	25 350	27 081	27 759	28 499	**28 603**
Cameroon	10 062	9 984	10 241	10 608	10 288	10 607	10 596	**10 681**
Côte d'Ivoire	13 601	13 715	13 681	14 636	15 346	15 488	16 173	**16 356**
Benin	5 975	6 273	6 429	6 820	7 051	7 104	7 277	**7 565**
Guinea-Conakry	4 561	4 897	4 693	4 827	4 546	3 427	3 338	**3 189**
Congo-Brazzaville	3 135	3 132	3 230	3 195	3 204	3 288	3 336	**3 308**
Liberia	1 731	1 768	1 845	1 875	1 875	1 928	1 975	**2 047**
Guinea-Bissau	865	889	837	801	811	821	838	**870**
MENA	65 323	65 410	65 154	65 702	66 541	66 864	66 992	**67 821**
Iran (joint venture)[a]	49 206	49 417	49 569	50 385	51 353	51 814	52 114	**52 747**
Sudan	9 817	9 714	9 433	9 213	9 215	9 109	8 839	**9 042**
Afghanistan	6 299	6 279	6 152	6 105	5 973	5 941	6 040	**6 033**
Total subscribers	267 750	266 728	266 809	272 398	276 206	281 620	284 900	**289 112**

Note: [a] Includes Wimax.

Source: MTN (2023a) Financial Reports Data Sheets, FY 2022.

Table 10 Revenue (in Millions)

Country	2022	2021[a]	Reported (%)	Constant currency[b] (%)
South Africa	**50 640**	48 716	3.9	3.9
Nigeria	**77 260**	60 050	28.7	21.7
SEA	**19 342**	16 955	14.1	11.6
Uganda	**10 126**	8 549	18.4	11.0
Rwanda	**3 527**	2 775	27.1	19.1
Zambia	**3 316**	2 429	36.5	7.1
South Sudan	**2 325**	2 998	(22.4)	19.3
Business Group	**48**	204	(76.5)	(76.7)
WECA	**48 279**	48 005	0.6	12.7
Ghana	**18 031**	19 187	(6.0)	28.6
Cameroon	**7 727**	7 244	6.7	8.7
Côte d'Ivoire	**8 918**	8 903	0.2	2.1
Benin	**6 768**	6 281	7.8	9.8
Guinea-Conakry	**1 668**	1 505	10.8	(10.5)
Congo-Brazzaville	**3 385**	3 379	0.2	2.1
Liberia	**1 507**	1 098	37.2	24.2
Guinea-Bissau	**275**	408	(32.6)	(31.1)
MENA	**6 212**	6 550	(5.2)	68.2
Sudan	**4 032**	2 226	81.1	157.5
Afghanistan	**2 180**	2 092	4.2	2.0
Syria	–	308	(100.0)	0.0
Yemen	–	1 924	(100.0)	0.0
Joint ventures		6 725	27.3	30.1
Iran	**8 564**	1 147	(4.3)	(3.7)
Botswana	**1 098**	445	2.0	2.0
Eswatini	**454**	(8 317)		
Equity-accounting exclusion	**(10 116)**			
Head office, GlobalConnect and eliminations	**2 786**	1 383		
Total	**204 519**	181 659	12.6	14.8
Hyperinflation	**2 484**	(13)		
Total including hyperinflation	**207 003**	181 646	14.0	14.8

Note: [a] Reported numbers are inclusive of MTN Syria and MTN Yemen up to the date of deconsolidation effective February 2021 and October 2021 respectively; [b] Constant currency excludes Yemen and Syria.

Source: MTN (2023) Financial Reports Data Sheets, FY 2022.

Table 11 Vodacom Group Ltd Profile (as of 30 September 2023)

Country	Population (mil)	Customers (thousands)	Licence expiry	Real GDP (USD)	Ownership (%)
South Africa	61.3	47256	2041	0.6	100
Democratic Republic of Congo (DRC)	102.3	21785	2028/2032/2038[a]	6.5	51
Tanzania	67.4	18502	2031	5.5	75
Mozambique	33.9	11867	2038	6.5	85
Lesotho	2.3	1531	2036	0.7	85
Kenya (Safaricom)	55.1	44115	2032/2024/2026[b]	5.2	34.9[c]
Ethiopia (Safaricom)	126.5	4125	2036	6	5.7[d]
Egypt	112.7	47022	2031	3.7	55

Note: [a] 2028 (2G licence), 2032 (3G licence), 2038 (4G licence); [b] 2032 (3G licence), 2024 (2G licence), 2026 (4G licence); [c] Vodacom Group Limited owns 87.5 per cent of Vodafone Kenya Ltd, which in turn holds 39.93 per cent of Safaricom Plc, giving Vodacom part ownership of Safaricom; [d] The Group has an effective interest of 5.7 per cent in Ethiopia, excluding its indirect interest via its shareholding in Safaricom PLC. Additionally, the Group has indirect exposure to Ethiopia through Safaricom PLC's 51.7 per cent effective interest in Ethiopia.

Source: Vodacom Group Factsheet, 2023.

Key ICT indicators (network coverage rates), 2015–22

Table 12 Total Population Covered by a Mobile-Cellular Network (%)

	2015	2016	2017	2018	2019	2020	2021	2022
Africa	86.6	87.5	87.6	89.0	88.7	90.1	90.8	91.5
Americas	95.8	95.9	96.2	96.1	96.0	95.9	96.0	96.0
Arab States	93.3	94.1	95.5	95.0	96.3	96.7	96.7	96.4
Asia Pacific	96.1	96.6	97.7	97.9	98.6	98.5	98.6	98.8
CIS	90.3	92.4	93.5	93.6	98.6	98.6	98.9	99.0
Europe	99.5	99.6	99.7	99.7	99.6	99.8	99.8	99.8

Source: ITU World Telecommunication/ICT Indicators database, accessed 23 November 2023. https://datahub.itu.int/data/.

Table 13 Total Urban Population Covered by a Mobile-Cellular Network (%)

	2015	2016	2017	2018	2019	2020	2021	2022
Africa	99.4	100.0	100.0	100.0	100.0	100.0	100.0	100.0
Americas	99.9	99.9	100.0	100.0	100.0	100.0	100.0	100.0
Arab States	98.9	99.6	100.0	99.9	99.9	99.9	100.0	100.0
Asia-Pacific	100.0	100.0	100.0	100.0	100.0	100.0	100.0	100.0
CIS	100.0	100.0	100.0	100.0	100.0	100.0	100.0	100.0
Europe	100.0	100.0	100.0	100.0	100.0	100.0	100.0	100.0

Source: ITU World Telecommunication/ICT Indicators database, accessed 23 November 2023.

Table 14 Total Rural Population Covered by a Mobile-Cellular Network (%)

	2015	2016	2017	2018	2019	2020	2021	2022
Africa	78.5	79.5	79.4	81.6	81.0	83.2	84.2	85.3
Americas	78.2	78.6	79.5	79.3	76.2	77.7	78.0	78.2
Arab States	85.4	86.3	89.2	87.9	88.2	91.9	92.0	92.1
Asia-Pacific	92.7	93.5	95.5	95.9	97.2	97.1	97.1	97.6
CIS	71.6	77.9	81.1	81.2	95.8	95.9	96.7	96.9
Europe	98.0	98.5	98.7	98.6	98.6	99.0	99.0	99.0

Source: ITU World Telecommunication/ICT Indicators database, accessed 23 November 2023.

Table 15 Black Economic Empowerment in Telecommunications (EXCO Members), 2018–22

	Telecoms employment – Black top management (EXCO Members)	Telecoms employment – top management (EXCO Members)	Telecoms employment– Black top female management (EXCO Members)
2018	68	255	29
2019	81	287	28
2020	83	232	37
2021	104	257	48
2022	81	299	40

Source: ICASA (2023: 39).

Table 16 B-BBEE Procurement Spend to Suppliers, 2018–22

	2018 (ZAR)	**2019 (ZAR)**	**2020 (ZAR)**	**2021 (ZAR)**	**2022 (ZAR)**
Procurement spend from all suppliers	137 963 97 264	153 709 327 406	99 474 488 024	124 166 572 439	166 088 114 064
Procurement spend based on the B-BBEE ranking levels	91 387 499 397	125 580 981 745	82 629 651 919	105 744 926 535	140 656 25 7 045

Source: ICASA (2023: 28).

Appendix D

MTN timeline

1994	• MTN is founded as M-Cell with assistance from the South African government, rolling out South Africa's first GSM payphones. • The first call is made on the MTN network.
1995	• M-Cell becomes a public company with 25 per cent of MTN Holdings and 60 per cent of M-Tel. • MTN creates a community services division to increase teledensity in disadvantaged areas. • MTN offers free calls to emergency services for the first time in South Africa.
1996	• MTN acquires South Africa's biggest service provider in M-Tel, creating the blueprint for network-owned service provision. • MTN develops a telemetric app for vehicle tracking and theft recovery via GSM. • MTN launches the prepaid solution PayAsYouGo, making cell phones accessible to unbanked South Africans. • MTN makes short message service (SMS) available to its customers.
1997	• MTN acquires services provider Transtel Cellular. • MTN begins its international expansion drive.
1998	• MTN International is established. • MTN becomes the first cellular network in Africa to receive ISO 14001 rating for environmental management. • MTN offers South Africa's first two-way PC-to-cellphone messaging app as a free and international service. • MTN wins a license to operate in Rwanda and Uganda.
2001	• Johnnic Holdings becomes the controlling shareholder • MTN acquires first-tier ISP Citec to enhance data services offering to the business sector. • MTN becomes a tier-one Internet Service Provider (ISP), trading as MTN Network Solutions. • MTN SA Foundation is established to manage MTN's multimillion rand CSI fund.

2002 • MTN installs a solar-powered payphone on Lake Victoria for local fishermen.
• MTN is the first operator in South Africa to offer General Packet Radio Services (GPRS).
• MTN Business is launched to deliver dedicated connectivity solutions and services to the corporate sector.

2003 • MTN VillagePhone is launched in association with Grameen Foundation, creating opportunities for entrepreneurs in Uganda's rural areas.
• MTN makes South Africa's first video call using its new 3G-enabled network.

2004 • MTN wins a license to operate in Cyprus, launching the country's first 4G network in 2015.

2005 • MTN is one of the first operators in Africa to bring 3G to the market

2006 • MTN reaches 40 million subscribers across the Group.

2007 • Online banking services and mobile TV through MultiChoice are introduced in South Africa.

2008 • MTN pioneers instant messaging in South Africa, enabling groups to chat simultaneously in real time.
• May: Bharti Airtel explores the possibility of buying MTN Group. Reliance Communications also considers a potential combination of businesses. Discussions with both companies end in July.
• June: MTN Group acquires Verizon Business South Africa.

2009 • MTN hits the 100 million subscribers mark in Africa and the Middle East.
• MTN launches mobile money in Benin, Ghana, Cameroon, Ivory Coast, Uganda, and Yemen.
• MTN Business deploys server virtualization offering in South Africa.
• February: The acquisition of Verizon Business South Africa is completed.
• June: MTN Group's subsidiary merges with Belgacom International Carrier Services (BICS).

2012 • MTN launches a commercial LTE network in South Africa, the largest rollout of its kind on the continent.
• October: MTN announces a partnership with Afrihost to provide DSL Broadband services in Africa.
• November: Shanduka Group acquires a minority stake in MTN Group's Nigeria business for $335 million.

2013 • MTN partners with Rocket Internet to enhance internet in Africa and the Middle East.

2015 • MTN acquires multimedia and telecoms company Smart Village, starting the rollout of fibre-to-the-home across South Africa.

2018 • MTN partners with KaiOS Technologies, China Mobile and UNISOC to launch Africa's first 3G smart feature phone.
• 31 May : MTN Ghana launches its initial public offering (IPO).

2019	• MTN achieves the completion of the largest BEE transaction in the telecommunications sector. • 31 July: The IPO for MTN Ghana closes, offering 35 per cent of its equity. MTN Ghana deploys 4G LTE mobile services as part of the agreement with Ghana's National Communications Authority.

Source: M-Cell (2001); MTN (2020).

Vodacom timeline

1993	Awarded a license to operate a GSM cellular network in South Africa.
1995	Vodacom is launched in Tanzania.
1996	Vodacom becomes the first to launch prepaid services worldwide.
1997	Vodacom expands its operations by launching in Mozambique.
2002	Vodacom launches a state-of-the-art 3G HSDPA network in South Africa.
2003	Vodacom Business is launched.
2006	Vodacom is listed on the JSE (18 May).
2008	Vodacom achieves a milestone by being the first to launch 4G in South Africa.
2009	Vodacom acquires 35 per cent of Safaricom in Kenya.
2012	Vodacom becomes the first telco in Africa to launch a commercial 5G network in Lesotho.
2017	Vodacom extends its reach by launching in Lesotho.
2018	A landmark agreement is made with Alipay to drive digital and financial inclusion.
2019	Vodacom completes the largest BEE (Black Economic Empowerment) transaction in the telecommunications sector.
2020	VodaPay and M-Pesa superapps are launched.
2021	Vodacom launches M-Pesa in Ethiopia.
2022	The acquisition of Vodafone Egypt is concluded.

Source: Adapted from Vodacom (2023).

Appendix E

Glossary

ADSL (Asymmetric Digital Subscriber Line) A type of DSL technology that allows faster download speeds than upload speeds.

Affirmative Action Policies aimed at increasing opportunities for historically disadvantaged groups.

Airtime The credit or minutes available for making calls or using mobile services.

Apartheid The historical policy of racial segregation and discrimination in South Africa.

Bandwidth The range of frequencies a communication channel can carry.

BBBEE (Broad-Based Black Economic Empowerment) South African government's policy promoting economic transformation and inclusivity

BEE Commission Regulatory body overseeing BEE compliance and reporting

Black Empowerment Charter Industry-specific guidelines for implementing BEE principles.

Black Industrialist Program Initiative supporting the growth of Black-owned businesses in key sectors.

Black-owned Enterprise A business where Black individuals have a significant ownership stake.

Broadband High-capacity transmission that supports multiple signals simultaneously.

Broadband High-speed internet access that allows for the transmission of large amounts of data.

Cellular Network A network of interconnected base stations for mobile communication.

Connectivity The state of being connected to a network, typically referring to internet or telecommunications networks.

Data Bundle A prepaid or postpaid package that includes a specific amount of data for internet access.

Digital Divide The gap between those who have access to modern information and communication technology and those who do not.

Digital Literacy The ability to access, understand and use digital technologies including mobile phones.

DSL (Digital Subscriber Line) A technology that provides high-speed internet access over traditional copper telephone lines.

FCC (Federal Communications Commission) The regulatory agency overseeing communications in the United States.

Fibre Optic Cable A type of cable that transmits data using light pulses through optical fibres, providing high-speed internet connectivity.

Fixed-Line Telecommunications Traditional landline services provided by Telkom for voice communication.

GSM (Global System for Mobile Communications) Standard for mobile communication.

ICT (Information and Communication Technology) Technologies used to manipulate and communicate information, including computers, the internet, and telecommunications.

Inclusive Procurement BEE requirement for businesses to procure goods and services from Black-owned enterprises.

Interconnection The linking of different telecommunications networks to enable them to communicate with each other.

JSE (Johannesburg Stock Exchange) South Africa's primary stock exchange.

Mobile Broadband High-speed internet access available through mobile networks.

Mobile Network Operator (MNO) A company that provides wireless communication services to mobile device users.

Mobile Telephony The use of wireless technology to transmit voice and data communication using mobile devices.

Mobile Virtual Network Operator (MVNO) A company that provides mobile services using the infrastructure of other mobile operators.

MTN Group The multinational telecommunications group overseeing MTN's operations globally.

National Roaming The ability of a mobile subscriber to use the services of another operator when outside the coverage area of their own operator.

Outsourcing Contracting out certain business functions to external service providers.

Ownership Scorecard BEE measurement of the extent to which Black individuals own and control a business.

Preferential Procurement BEE requirement for businesses to prioritize Black-owned suppliers.

Prepaid A mobile service plan where users pay for services in advance.

Privatization Transfer of ownership and control of government-owned enterprises to the private sector.

Public Access Terminal A facility providing public access to computers and the internet.

Public-Private Partnership (PPP) Collaboration between government and private entities for public projects.

RDP (Reconstruction and Development Programme) South African government initiative for social and economic development.

SMS (Short Message Service) A mobile phone text messaging service

Spectrum The range of radio frequencies used for wireless communication, including mobile networks.

Spectrum Allocation The assignment of specific radio frequencies to different services or users.

Subscriber Identity Module (SIM) A small card inserted into mobile phones, providing identification and connectivity to the mobile network.

Telecom Monopoly A situation where a single company dominates the telecommunications market.

Telecom Policy Government policies governing the telecommunications industry.

Telecom Regulatory Authority A government agency responsible for regulating the telecommunications industry.

Telecom Regulatory Framework The legal and regulatory structure governing telecommunications.

Telecommunication Infrastructure The physical and technical components that enable the transmission of information over a distance.

Telkom Group The parent company overseeing Telkom's operations in various sectors.

Telkom Integrated Report Telkom's annual report providing a comprehensive overview of its performance and activities.

Telkom Mobile Telkom's mobile service division, providing cellular services to subscribers.

Telkom Payphone Public telephones operated by Telkom that accept various payment methods.

Universal Access Ensuring that everyone has equal opportunities to access and use information and communication technologies.

Universal Service The concept that all citizens should have access to essential telecommunication services.

Universal Service Fund A fund established to support providing essential telecommunications services to all citizens, especially in underserved areas.

Vodafone Group Vodacom's association with the global telecommunications group that holds a significant stake in the company.

Wireless Fidelity (Wi-Fi) Technology that allows electronic devices to exchange data wirelessly over a local area network.

References

Access Now. 2022. ‘The Return of Digital Authoritarianism’. Available at: https://www.accessnow.org/wp-content/uploads/2022/05/2021-KIO-Report-May-24-2022.pdf (Accessed 7 August 2023).

Access Now. 2023. ‘Internet Shutdowns in 2022’. Available at: https://www.accessnow.org/wp-content/uploads/2023/05/2022-KIO-Report-EECA.pdf (Accessed 7 August 2023).

Acharya, U. 2013. ‘The Future of Human Development: The Right to Survive as a Fundamental Element of the Right to Development’. *Denver Journal of International Law & Policy*, 42, 345.

Adam, H., Slabbert, F. V., & Moodley, K. 1997. *Comrades in Business: Post-Liveration Politics in South Africa*. Cape Town: Tafelberg.

Addeh, E. 2017. ‘Hoodlums Planning to Attack MTN Facilities in Bayelsa, NSCDC Warns’. *THISDAYLIVE*. Available at: https://www.thisdaylive.com/index.php/2017/02/28/hoodlums-planning-to-attack-mtn-facilities-in-bayelsa-nscdc-warns?amp=1 (Accessed 4 August 2023).

Addison, C. 1947. ‘Reference: Telegram from Secretary of State for Commonwealth Relations, London, Participation of Outer Mongolia in International Telecommunications Union’ [South African National Archives]. Circular D.WO.783, Confidential (DOM), 22 September 1947.

Adebajo, A. 2010. *The Curse of Berlin: Africa after the Cold War*. Preface by Ali A. Mazrui. London: Hurst & Co.

Adebajo, A. 2017. *The Eagle and the Springbok: Essays on Nigeria and South Africa*. London: Routledge.

Africa & Middle East Telecom-Week. n.d. ‘Southern Africa Mobile Operators: Overview of Vodacom and Virgin’. http://www.africantelecomsnews.com/resources/South_African_mobile_market.shtml (Accessed 1 August 2015).

Africa Progress Panel. 2012. *Jobs, Justice, Equity: Seizing Opportunities in Times of Global Change*. Geneva: Africa Progress Panel.

‘Africa’s New Number One’. 2015. *The Economist*. Available at: http://www.economist.com/news/leaders/21600685-nigerias-suddenly-supersized-economy-indeed-wonder-so-are-its-still-huge (Accessed 1 March 2016).

Ajayi, G., Salawu, R., and Raji, T. 1999. 'Nigeria: After a Century of Telecommunications, What Next?', in E. M. Noam (ed.), *Telecommunications in Africa*. Oxford: Oxford University Press on Demand, pp. 163–77.

Aker, Jenny C., and Mbiti, Isaac M. 2010. 'Mobile Phones and Economic Development in Africa'. *Journal of Economic Perspectives*, 24(3), 207–32.

Akinola, K. 2014. 'Technology Times Founder: Mobile Networks Are the New "One-Stop-Shop" Businesses'. http://technologytimes.ng/technology-times-foundermobile-phone-companies-has-now-become-business-service-provider-in-nigeria/ (Accessed 1 August 2015).

Akwaja, C. 2013. 'Nigeria: MTN Acquires 33.3 Percent Equity in Jumia's Parent Company'. http://allafrica.com/stories/201312170283.html (Accessed 3 February 2017).

Allison, S. 2014. 'South Africa to Rwanda: Don't Touch Us on Our Sovereignty!' https://www.dailymaverick.co.za/article/2014-03-10-south-africa-to-rwanda-dont-touch-us-on-our-sovereignty/#.WKmiY7Fh1Z0 (Accessed 3 February 2017).

Anand, S. 2008. Merger Talks Fail between Reliance and MTN'. http://www.financeasia.com/News/117285,merger-talks-fail-between-reliance-and-mtn.aspx (Accessed 6 June 2016).

ANC. 2007. '52nd National Conference: Resolutions'. http://www.anc.org.za/show.php?id=2536 (Accessed 30 March 2016).

Anwar, S. T. 2003. 'CASES Vodafone and the Wireless Industry: A Case in Market Expansion and Global Strategy'. *Journal of Business & Industrial Marketing*, 18(3), 270–88.

AP. 2023. 'MTN Nigeria, Shoprite Close South Africa Stores because of Increasing Violence'. CGTN Africa. Available at: https://africa.cgtn.com/mtn-nigeria-shoprite-close-south-africa-stores-because-of-increasing-violence/ (Accessed 4 August 2023).

Austin, B. 2001. *Wireless in the Boer War*. South African Military History Society. Available at: http://samilitaryhistory.org/diaries/wirelessabw.html (Accessed 7 August 2023).

Barkan, J. 2013. *Corporate Sovereignty: Law and Government under Capitalism*. Minneapolis: University of Minnesota Press.

Battersby, J. D. 1988, 'South Africa Quells Coup Attempt in a Homeland'. *Special to the New York Times*, 11 February.

BBC. 2010. 'Mozambique Reverses Bread Price Rise in Wake of Riots'. http://www.bbc.com/news/business-11216009 (Accessed 1 January 2017).

BBC News. 2017. 'Why Has Cameroon Blocked the Internet?' http://www.bbc.com/news/world-africa-38895541 (Accessed 8 February 2017).

BBC News Staff. 2016. 'Nigeria's Buhari Says MTN Fuelled Boko Haram Insurgency'. http://www.bbc.com/news/business-35755298 (Accessed 31 March 2016).

Bendix, Sonia. 2010. *Industrial Relations in South Africa*. Juta, 2010.

Bharti Airtel Ltd. 2009. 'Media Statement from Bharti Airtel Limited'. http://www.airtel.in/about-bharti/media-centre/bharti-airtel-news/corporate/pg-media_statement_from_bharti_airtel_limited (Accessed 4 June 2016).

Bloomberg, B. 2023. 'Africa's Biggest Wireless Carrier MTN Reviews Nigerian Tax Demand for $72.6 Million'. *Daily Maverick*. Available at: https://www.dailymaverick.co.za/article/2023-10-25-africas-biggest-wireless-carrier-mtn-reviews-nigerian-tax-demand-for-72-6-million/ (Accessed 25 October 2023).

Bond, P. 2004. 'South African Subimperialism and NEPAD'. IDEAs, Ethiopian Economic Association and the Council for the Development of Social Science Research in Africa Conference on The Agrarian Constraint and Poverty Reduction. https://ccs.ukzn.ac.za/files/Bond%20SA%20subimperialism.pdf (Accessed 18 March 2024).

Bond, P. 2006. *Looting Africa: The Economics of Exploitation*. London: Bloomsbury Publishing.

Bond, P., and Kapuya, T. 2006. 'Arrogant, Disrespectful, Aloof and Careless: South African Corporations in Africa'. *OSISA OPENSPACE*, 1(4), 32–3.

Booth, D., and Golooba-Mutebi, F. 2012. 'Developmental Patrimonialism? The Case of Rwanda'. *African Affairs*, 111(444), 379–403.

Bouch, R. 1993. 'Glen Grey before Cecil Rhodes: How a Crisis of Local Colonial Authority Led to the Glen Grey Act of 1894'. *Canadian Journal of African Studies/Revue Canadienne Des Études Africaines*, 27(1), 1–24.

Bowman, A. 2020. 'Parastatals and Economic Transformation in South Africa: The Political Economy of the Eskom Crisis'. *African Affairs*, 119(476), 395–431.

Bowman, Gilfillan. 2004. 'Foreign Companies and BEE'. http://www.bowman.co.za/News-Blog/Blog/FOREIGN-COMPANIES-AND-BEE (Accessed 18 February 2015).

BrandSA. 2013. 'Black Economic Empowerment'. https://www.brandsouthafrica.com/investments-immigration/business/trends/empowerment/bee (Accessed 14 February 2017).

Brederode, W. 2023. *How Much Tax MTN Pays – MyBroadband*. Available at: https://mybroadband.co.za/news/telecoms/490097-how-much-tax-mtn-pays.html (Accessed 25 October 2023).

Burger, R., Steenkamp, C., van de Berg, S., and Zoch, A. 2013. *The Emergent SA Middle Class*. Research on Socio-Economic Policy (RESEP). Available at: https://resep.sun.ac.za/wp-content/uploads/2013/10/The-emergent-SA-middle-class_.pdf (Accessed 30 October 2023).

Business Day. 1993a. 1 April, p. 8 (online NewsBank).

Business Day. 1993b. 18 June, p. 3 (online NewsBank).

Business Monitor International. 2009. *South African Telecommunications Report Quarter 2 2009*. London: Business Monitor International.

Business Standard. 2009. 'Bharti-MTN Deal Up against Affirmative Action Policy'. http://business.rediff.com/report/2009/sep/15/bharti-mtn-talks-south-african-govt-talks-tough.htm (Accessed 3 June 2016).

Cairns, P. 2010. 'The Investment Case – MTN Group'. http://www.moneyweb.co.za/archive/the-investment-case-mtn-group/ (Accessed 1 August 2015).

Calvo, A. G. 2014. *The Role of StateFirm Relationships in Fostering Competitiveness: Telefonica's Upgrading*. Available at: https://www.carloalberto.org/wp-content/uploads/2018/11/no.397.pdf (Accessed 1 September 2023).

Carlos, A. M., and Nicholas, S. 1988. '"Giants of an Earlier Capitalism": The Chartered Trading Companies as Modern Multinationals'. *Business History Review*, 62(3), 398–419.

Carey, J. W. 1992. *Communication as Culture: Essays on Media and Society* (Vol. 1). London: Routledge, 201.

Carney, B. T. 2008. *South Africa: Profile of Total ICT Spending and ICT Household Spending by Income Segment*. World Resource Institute. Available at: http://www.wri.org/resources/charts-graphs/south-africa-profile-total-ict-spending-and-ict-household-spending-income (Accessed 1 September 2014).

Carvajal, D. 2010. 'Hunting for Liberia's Missing Millions'. *New York Times*, A6, 30 May.

Cave, M., Genakos, C., and Valletti, T. 2019. 'The European Framework for Regulating Telecommunications: A 25-Year Appraisal'. *Review of Industrial Organization*, 55, 47–62.

Cawe, A. 2020. 'Gambling on the Share Price: Beyond Tenderpreneurship: Rethinking Black Business and Economic Empowerment', in A. Cawe and K. Mabasa (eds), *Beyond Tenderpreneurship: Rethinking Black Business and Economic Empowerment*. Johannesburg: Mapungubwe Institute for Strategic Reflection, 227–55.

Cawston, G., and Keane, A. 1896. *The Early Chartered Companies: A.D 1296–1858*. London: Edward Arnold.

Chandhoke, N. 1986. 'Crisis of Settler Hegemony in South Africa'. *Economic and Political Weekly*, 21(7), 291–4.

Channels Television. 2012. 'Mend Threatens to Attack MTN Facilities in Niger Delta'. Available at: https://www.channelstv.com/2012/04/15/mend-threatens-to-attack-mtn-facilities-in-niger-delta/ (Accessed 4 August 2023).

Charles-Iyoha, C. 2008. 'Comparative African Perspectives on the Trade Relations and Practices of Transnational Telecommunications Corporations of China and South Africa in Nigeria', in *Strengthening the Civil Society Perspective Series II: China and Other Emerging Powers in Africa*. Cape Town: Pambazuka Press, 30–65.

Charles-Iyoha, C. 2010. 'Mobile Telephony: Closing the Gap', in S. Ekine (ed.), *SMS Uprising: Mobile Phone Activism in Africa*. Cape Town: Pambazuka Press, 116–23.

Chinje, N. B. 2008. 'The Economic Impact of MTN's involvement in Cameroon'. MBA Dissertation. University of Stellenbosch. CISCO 2017.

Cisco Visual Networking Index. 2016. 'Global Mobile Data Traffic Forecast, 2016–2021'. Available at: https://www.cisco.com/c/dam/m/en_us/solutions/service-provider/vni-forecast-highlights/pdf/Global_2021_Forecast_Highlights.pd (Accessed 9 February 2017).

Cissé, D. 2012. 'Chinese Telecom Companies Foray into Africa'. *African East-Asian Affairs: The China Monitor*, 69.

Cobham, A., and Janský, P. 2018. 'Global Distribution of Revenue Loss from Corporate Tax Avoidance: Re-estimation and Country Results'. *Journal of International Development*, 30, 206–32.

Cochran, E. S. 2000–1. 'The Pivotal State: Post-Apartheid South Africa'. *Parameters*, 30(4): 83–101.

Codding Jr, G. A. 1989. 'Financing Development Assistance in the ITU'. *Telecommunications Policy*, 13(1), 13–24.

Codding Jr., G. A. 1991. 'Evolution of the ITU'. *Telecommunications Policy*, 15(4), 271–85.

Codding Jr., G. A. 1994. 'The International Telecommunications Union: 130 Years of Telecommunications Regulation'. *Denver. Journal of International Law & Policy*, 23, 501.

Codding, G. A 1952. *The International Telecommunication Union: An Experiment in International Cooperation*. Leiden: E. J. Brill.

Coetzer, P. 2008. 'Low Income Does Not Mean No Income' a Vodacom Psitek Facts FV – the Bop Lab (Southern Africa). Reciprocity. https://www.b4dpathfinder.org/documents/Reciprocity_BoP-FactSheet_Vodacom.pdf (Accessed 18 March 2024).

Cogburn, D. L. 1998. 'Globalization and State Autonomy in the Information Age Telecommunications Sector Restructuring in South Africa'. *Journal of International Affairs*, 51(2), 583–604.

Cohen, T. 2003. 'Rethinking (Reluctant) Capture: South African Telecommunications and the Impact of Regulation'. *Journal of African Law*, 47(1), 65–87. http://www.jstor.org/stable/30038548 (Accessed 2 March 2016).

Cole, M. 2015. *Is South Africa Operating in a Safe and Just Space? Using the Doughnut Model to Explore Environmental Sustainability and Social Justice.* Oxford: Oxfam. https://oxfamilibrary.openrepository.com/bitstream/handle/10546/555842/rr-south-african-doughnut-sustainability-social-justice-280515-en.pdf;jsessionid=00D390B8E98113DB689D0AD67C9497BE?sequence=7 (Accessed 1 May 2016).

Comaroff, J., and Comaroff, J. L. 1992. 'Home-Made Hegemony: Modernity, Dosmesticity and Colonialism in South Africa', in K. Tranberg (ed.), *African Encounters with Domesticity.* New Brunswick: Rutgers University Press, 37–74.

Congchiglia, A. 2007. 'South Africa and Its Lusopone Neighbours: Angola and Mozambique', in A. Adebajo, A. Adedeji and C. Landsberg (eds), *South Africa in Africa: The Post-Apartheid Era*. Scottsville: University of KwaZulu-Natal Press, 236–52.

Constitution of the Republic of South Africa. 1996: 108. https://www.justice.gov.za/legislation/constitution/saconstitution-web-eng.pdf (Accessed 18 March 2024).

'Cosatu Criticise South Africa's IMF Donation'. 2012. *Mail & Guardian Online.* Available at: http://mg.co.za/article/2012-06-19-cosatu-criticise-south-africas-imf-donation (Accessed 19 June 2012).

Cowhey, P. F. 1990. 'The International Telecommunications Regime: The Politics of Regimes for High Technology'. *International Organization*, 442, 169–99.

Cox, R. 1976. 'Labour and the Multinationals'. *Foreign Affairs*, 54, 344–65.

Cox, R. W., and Jacobson, H. K. 1973. *The Anatomy of Influence: Decision Making in International Organization*. New Haven: Yale University Press.

Creamer, T. 2016. 'DTI Sets up Africa-Focused Trade Unit in Bid to Grow Exports Beyond R300bn'. *Creamer Media's Engineering News*, 15 July. Available at: http://www.engineeringnews.co.za/article/dti-sets-up-africa-focused-export-unit-in-bid-to-grow-exports-beyond-r300bn-2016-07-15/rep_id:4136 (Accessed 2 August 2016).

Cuervo-Cazurra, A., Inkpen, A., Musacchio, A. and Ramaswamy, K. 2014. 'Governments as Owners: State-Owned Multinational Companies'. *Journal of International Business Studies*, 45(8), 919–42.

Culpepper, P. 2015. 'Structural Power and Political Science in the Post-Crisis Era'. *Business and Politics*, 17(3), 391–409.

Curwen, P., and Whalley, J. 2004. *Telecommunications Strategy: Cases, Theory and Applications*. London: Routledge.

Cwele, S. 2018. 'SA: Siyabonga Cwele: Address by Minister of Telecommunications and Postal Services, During the 2018/19 Telecommunications and Postal Services Pre Budget Vote'. Media Briefing, Parliament, Cape Town, 17 May. Available at: http://www.polity.org.za/article/sa-siyabonga-cwele-address-by-minister-of-telecommunications-and-postal-services-during-the-201819-telecommunications-and-postal-services-pre-budget-vote-media-briefing-parliament-cape-town-17052018-2018-05-18 (Accessed 28 July 2023).

Dada, A., and Ubabukoh, O. 2015. 'We've So Far Paid N1.3tn to Govt – MTN Nigeria'. *Punch Online*, 5 March. Available at: http://www.punchng.com/business/business-economy/weve-so-far-paid-n1-3tn-to-govt-mtn-nigeria/ (Accessed 9 September 2015).

Dagada, R. 2012. 'Employing ICT to Leapfrog the South African Economy 51 into the Fourth Wave'. *Focus*, 66. https://hsf.org.za/publications/focus/focus-66/RDagada.pdf (Accessed 18 March 2024).

Dahir, A. L. 2016. 'The Number of Tech Hubs across Africa Has More Than Doubled in Less Than a Year'. *Quartz Africa*, 17 August. Available at: https://qz.com/759666/the-number-of-tech-hubs-across-africa-has-more-than-doubled-in-less-than-a-year/ (Accessed 20 January 2017).

Daniel, J., Lutchman, J. and Naidu, S. 2004. 'Post-Apartheid South Africa's Corporate Expansion into Africa'. *Review of African Political Economy*, 31(100), 343–78.

Davenport, J. 2013. *Digging Deep: A History of Mining in South Africa*. Cape Town: Jonathan Ball.

Davidson, N. 2012. 'Mapping and Effectively Restructuring Operator-Bank Relationships to Offer Mobile Money for the Unbanked'. GSMA. https://www.gsma.com/mobilefordevelopment/wp-content/uploads/2012/03/mappingandeffectivestructuringfinal2643.pdf (Accessed 18 March 2024).

Davies, R. 1992. 'Economic Growth in a Post-Apartheid South Africa: Its Significance for Relations with Other African Countries'. *Journal of Contemporary African Studies*, 11(1): 51–71.

de Deugd, N., and Hoen, H. 2010. *Dovetailing Economics and Political Science: A Paradigmatic Introduction to International Political Economy*. Assen: Uitgeverrij Van Gorcum.

De Klerk, F. 1990. 'Address by the State President at the Opening of the Second Session of the Ninth Parliament of the Republic of South Africa'. Cape Town. https://www.gov.za/news/address-state-president-mr-fw-de-klerk-dms-opening-second-session-ninth-parliament-republic (Accessed 18 March 2024).

de Kock, P. 2014. *South Africa Inc Series: African Market Entry Strategy*. Brand South Africa Research Note 2. Brand South Africa.

de Kock, P. 2015. 'Research & Nation Brand Performance Presentation at the Offices of SARCHI Chair for African Diplomacy and Foreign Policy'. University of Johannesburg, South Africa.

De Villiers, W. J., and South Africa. 1989. *Summarised Report on the Study by Dr W.j. De Villiers Concerning the Strategy Policy Control Structure and Organisation of Posts and Telecommunications*. Pretoria: Department of Posts and Telecommunications.

Deetz, S. 1996. 'Describing Differences in Approaches to Organization Science: Rethinking Burrell and Morgan and Their Legacy'. *Organization Science*, 7(2): 191–207.

Deloitte. 2012. 'What Is the Impact of Mobile Telephony on Economic Growth: A Report for the GSM Association'. GSMA. Available at: http://www.gsma.com/publicpolicy/wp-content/uploads/2012/11/gsma-deloitte-impact-mobile-telephony-economic growth.pdf (Accessed 25 January 2015).

Deloitte. 2014. 'The Future of Telecoms in Africa: The "blueprint for the brave"'. Available at: https://www2.deloitte.com/content/dam/Deloitte/fpc/Documents/secteurs/technologies-medias-et-telecommunications/deloitte_the-future-of-telecoms-in-africa_2014.pdf (Accessed 30 October 2023).

Demian, M. 2015. 'Have Protests in South Africa Nearly Doubled since 2010?' *Africa Check*, 22 July. Available at: https://africacheck.org/reports/have-protests-in-south-africa-nearly-doubled-since-2010/ (Accessed 21 October 2015).

Department of Foreign Affairs. 1996. *South African Foreign Policy: Discussion Document*. Pretoria: Department of Foreign Affairs.

Department of International Relations and Cooperation. 2016a. 'Minister Nkoana-Mashabane Concludes Successful Participation in the Middle East Peace Initiative Ministerial Meeting in Paris'. Department of International Relations and Cooperation, 3 June. Available at: http://www.dirco.gov.za/docs/2016/mepp0606.htm (Accessed 3 June 2016).

Department of International Relations and Cooperation. 2016b. 'Update on African Renaissance and International Cooperation Fund: Departmental Briefing'. Parliamentary Monitoring Group, 24 August. Available at: https://pmg.org.za/committee-meeting/23125/ (Accessed 20 January 2017).

Department of International Relations and Cooperation. n.d. *Strategic Plan 2013–2018*. Pretoria: DIRCO.

Department of Trade and Industry, South Africa. 2009. *South Africa: Investors Handbook*. Pretoria: Department of Trade and Industry.

Deyner, S. 2016. 'China's Scary Lesson to the World: Censoring the Internet Works'. *The Washington Post*, 23 May. Available at: https://www.washingtonpost.com/

world/asia_pacific/chinas-scary-lesson-to-the-world-censoring-the-internet-works/2016/05/23/413afe78-fff3-11e5-8bb1-f124a43f84dc_story.html?utm_term=.99bb94d7e059 (Accessed 3 January 2017).

Dhliwayo, J. 2013. 'A Brief History of Telecommunications in Africa'. *ICT Africa*, 21 August. Available at: http://www.ictafrica.info/FullNews.php?id=10568 (Accessed 28 May 2016).

Dike, M. C., and Rose, E. L. 2017. 'Internationalization of Mobile Telecommunications: A Systematic Literature Review'. *Review of International Business and Strategy*, 27(3), 308–21.

Diop, C. A. 1996. *Towards the African Renaissance: Essays in African culture & development, 1946–1960*. Trenton, NJ: Red Sea Press.

DO4Africa. 2022. 'Digital Literacy in Africa'. *Do4africa*. Available at: https://www.do4africa.org/en/digital-literacy-in-africa/ (Accessed 29 October 2023).

Dolan, D. 2012. 'MTN Slides on Iran Corruption Lawsuit'. *Reuters*, 30 March. Available at: http://www.reuters.com/article/2012/03/30/us-mtn-iran-shares-idUSBRE82T0N620120330 (Accessed 30 March 2012).

Dorward, D. C. 1969. 'The Development of the British Colonial Administration among the Tiv, 1900–1949'. *African Affairs*, 68(273), 316–33.

Doyle, M. 1986. *Empires*. Ithaca: Cornell University Press.

DPME. 2012. *Mid-Term Review of the Priorities of Government*. Republic of South Africa: Department of Performance Monitoring and Evaluation in the Presidency.

Drucker, P. F. 1960. *Concept of the Corporation*. Boston: Beacon Press.

Du Plessis, C. 2016. 'The African Peer Review Mechanism: An Idea Whose Time Has Come Again?'. *The Daily Maverik*, 5 February. Available at: https://www.dailymaverick.co.za/article/2016-02-05-the-african-peer-review-mechanism-an-idea-whose-time-has-come-again/#.wj2BUrFh3FQ (Accessed 10 February 2017).

Dugard, J. 1973. *The South West Africa/Namibia Dispute: Documents and Scholarly Writings on the Controversy between South Africa and the United Nations*, vol. 9. Berkeley: University of California Press.

Dumako, A. 2016. 'SA Obligated to Ensure SACU Members Benefit Economically: Zuma'. *SABC News*, 23 April. Available at: http://www.sabc.co.za/news/a/1e7968004c81495fb5e0fff73ffe3ce9/SA-obligated-to-ensure-Sacu-members-benefit-economically:-Zuma-20162304 (Accessed 23 May 2016).

Dunsby, A., and Chochan, A. 2012. 'Enabling Possibilities through Mobile Payments'. *Communications Review*, 17(1): 34–41.

Dzineza, G. 2014. Interview Held at the ISS Pretoria Office. (O. P. Ayodele, Interviewer).

Eagle, N. 2010. 'Economics and Power within the African Telecommunications Industry', in S. Ekine (ed.), *SMS Uprising: Mobile Phone Activism in Africa*. Cape Town: Pambazuka Press.

Ebegbulem, J. C. 2013. 'An Evaluation of Nigeria – South Africa Bilateral Relations'. *Journal of International Relations and Foreign Policy*, 1(1): 32–40.

Ebrahim, E. I. 2011. 'Draft Lecture by Deputy Minister of International Relations and Cooperation, Ebrahim I Ebrahim, on the Occasion of the Speakers Meeting at the South African Institute of International Affair'. SAIIA. Available at: http://www.saiia.org.za/images/stories/saiia/events/ebrahim_ebrahim_july_2011/spe_dep_min_ebrahim_saiia_22_july_2011.pdf (Accessed 5 June 2016).

Ebrahim, S. 2015. 'SA and Mozambique Cement Ties'. *IOL*, 23 October. Available at: http://beta.iol.co.za/news/africa/sa-and-mozambique-cement-ties-1934535 (Accessed 23 October 2015).

Economic Press Review in the Ohada Zone October 2005. 2005. 'Ohada Legis'. Available at: http://www.ohadalegis.com/anglais/revueacteconarchoctnov05gb.htm (Accessed 20 August 2015).

Edigheji, O. 2003. 'State-Society Relations in Post-Apartheid South Africa: The Challenges of Globalisation on Co-operative Governance', in G. Mhone and O. Edigheji (eds), *Governance in the New South Africa: The Challenges of Globalisation*. Cape Town: University of Cape Town Press, 69–113.

Editor. 2016. 'Nigeria Present Woes Shows Limit of Buhari's Economic Policies'. *Financial Watch*, 21 March. Available at: http://www.financialwatchngr.com/2016/03/21/nigeria-present-woes-shows-limit-of-buharis-economic-policies/ (Accessed 31 March 2016).

El Habiti, H. 2022. 'Why Africa's Youth Hold the Key to Its Development Potential, World Economic Forum'. Available at: https://www.weforum.org/agenda/2022/09/why-africa-youth-key-development-potential/ (Accessed 5 September 2023).

Encarnation, D. J., and Vachani, S. 1985. 'Foreign Ownership: When Hosts Change the Rules'. *Harvard Business Review*, 63(5): 152–60.

Engineering News. 2003. 'African Communication Network Receives a R2-Billion Investment'. *Creamer Media: Engineering News*, 24 October. Available at: http://www.engineeringnews.co.za/article/african-communication-network-receives-a-r2billion-investment-2003-10-24 (Accessed 4 June 2016).

Enns, C., and Bersaglio, B. 2015. 'Enclave Oil Development and the Rearticulation of Citizenship in Turkana, Kenya: Exploring "Crude Citizenship"'. *Geoforum*, 67, 78–88.

Ensor, L. 2015. 'Treasury Offers Tax Deduction to SA Firms in Africa'. *Business Day BDlive*, 19 October. Available at: http://www.bdlive.co.za/business/2015/10/19/treasury-offers-tax-deduction-to-sa-firms-in-africa (Accessed 20 October 2015).

Ericsson. 2015. 'Ericsson Mobility Report: Sub-Saharan Africa'. Ericsson. Available at: https://www.ericsson.com/res/docs/2015/mobility-report/emr-nov-2015-regional-report-sub-saharan-africa.pdf (Accessed 12 May 2016).

Ernest & Young. 2015. 'Ernest & Young Attractiveness Survey 2015'. https://assets.ey.com/content/dam/ey-sites/ey-com/en_za/topics/attractiveness/reports/ey-aar-2015-making-choices.pdf (Accessed 18 March 2024).

Esipisu, M. 2002. 'Businesses See Value in Nepad'. Fin24Archives, 23 July. Available at: http://www.fin24.com/Economy/Businesses-see-value-in-Nepad-20020723 (Accessed 5 June 2016).

Estache, A., and Goicoechea, A. 2023. 'World Bank. Research Database on Infrastructure Economic Performance (IEP) 1980–2004. Ref. WLD_2004_IEP_v01_M'. Available at: http://microdata.worldbank.org (Accessed 1 February 2023).

Europa: Summaries of EU Legislation. 2010. 'Treaty Establishing the European Coal and Steel Community Fact Sheet'. Available at: http://europa.eu/legislation_summaries/institutional_affairs/treaties/treaties_ecsc_en.htm (Accessed 2 August 2013).

Evans, O. 2018. Connecting the Poor: The Internet, Mobile Phones and Financial Inclusion in Africa. *Digital Policy, Regulation and Governance*, 20(6), 568–81.

Evans, P. 1997. 'The Eclipse of the State? Reflections on Stateless in an Era of Globalization'. *World Politics*, 50(1): 62–87.

Fabricius, P. 2021. 'A Fit for Purpose Foreign Policy: Integrating Investment, Growth and Diplomacy'. Discussion paper 012/2021. Available at: https://www.thebrenthurstfoundation.org/downloads/fit-for-purpose-foreign-policy-final.pdf (Accessed 2 September 2021).

Fabricius, P., and Faubet, P. 2010. 'Cellphones Run Riot'. *Sunday Tribune*, 5 September 2010.

Fairbrother, P. 1998. 'The Telecommunications Industry: Questions of State, Politics and Labour'. *Work Employment & Society*, 12, 769–80.

Fari, S., Balbi, G., and Richeri, G. 2012. 'A Common Technical Culture of Telegraphy: The Telegraph Union and the Significance of Technological Standardization 1865–1875', in *History of Electro-Technology Conference (HISTELCON), Third IEEE*, 1–5. https://ieeexplore.ieee.org/document/6487580 (Accessed 18 March 2024).

Federal Communications Commission. 2015. 'Types of Broadband Connections'. Available at: https://www.fcc.gov/general/types-broadband-connections (Accessed 12 May 2016).

Feld, W. J. 1980. *Multinational Corporations and U.N. Politics: The Quest for Codes of Conduct*. New York: Pergamon Press.

Ferroni, M. 2004. 'Regional Public Goods and the Regional Development Banks'. Available at: http://www.cepal.org/portofspain/noticias/noticias/8/14868/ferroni pub.pdf (Accessed 29 October 2015).

Fin24. 2002. 'Deal Values Vodacom at R30bn, Business'. Available at: https://www.news24.com/fin24/deal-values-vodacom-at-r30bn-20020829 (Accessed 1 August 2023).

Fine, B., and Rustomjee, Z. 1996. *The Political Economy of South Africa: From Gold Mining to Minerals Industrial Complex*. Boulder, CO: Westview.

Flemes, D. 2007. 'Conceptualizing Regional Power in International Relations: lessons from the South African case'. GIGA Working Papers 53. http://dx.doi.org/10.2139/ssrn.1000123 (Accessed 18 March 2024).

Flemes, D., and Nolte, D. 2010. 'Introduction', in D. Flemes (ed.), *Regional Leadership in the Global System: Ideas, Interests and Strategies of Regional Powers*. Farnham: Ashgate, 1–14.

Fleshman, M. 2010. 'A "Wind of Change" that Transformed the Continent'. *Africa Renewal*, 28–30.

Ford, J. 2014. 'Engaging the Private Sector in Post-Conflict Recovery: Perspectives for SADPA'. ISS Paper 269. https://www.files.ethz.ch/isn/184710/Paper269-9Oct2014.pdf (Accessed 18 March 2024).

Fordred, L. L. 1997. 'Wireless in the Second Anglo Boer War 1899–1902'. *Transactions-South African Institute of Electrical Engineers*, 88(3), 61–71.

Foster, V., and Briceno-Garmendia, C. 2010. *Africa's Infrastructure: A Time for Transformation*. Washington, DC: World Bank and Infrastructure Consortium for Africa.

Fourie, J. 2006. 'The South African Poor White Problem in the Early 20th Century: Lessons for Poverty Today, Stellenbosch Economic Working Papers: 14/06'. Available at: https://resep.sun.ac.za/wp-content/uploads/2017/10/wp-14-2006.pdf (Accessed 26 July 2023).

Francis, D. 2016. 'Top 25'. *Foreign Policy*, 15 March. Available at: http://foreignpolicy.com/2016/03/15/these-25-companies-are-more-powerful-than-many-countries-multinational-corporate-wealth-power/ (Accessed 17 May 2016).

Francis, J. 2014. 'Talk Is Cheap'. *ITWeb*, 23 April. Available at: http://www.itweb.co.za/mobilesite/index.php?option=com_content&view=article&id=134000 (Accessed 1 September 2015).

Frankel, G. 1985. 'South Africa Imposes State of Emergency'. *The Washington Post*. Available at: https://www.washingtonpost.com/archive/politics/1985/07/21/south-africa-imposes-state-of-emergency/d125801e-fb8e-4069-9c7b-3626daf02cc5/ (Accessed 25 July 2023).

Front-Line States. n.d. 'South Africa History'. Available at: http://www.sahistory.org.za/organisations/frontline-states (Accessed 30 June 2012).

Gagliardone, I. 2016. *The Politics of Technology in Africa*. Cambridge: Cambridge University Press.

Games, D. 2003. *The Experience of South African Business in Africa Firms Doing Business in Africa: A Preliminary Survey and Analysis*. Johannesburg: South African Institute of International Affairs.

Games, D. 2010. 'Emerging Commercial Rivalries in Africa: A View from South Africa'. SAIIA Policy Briefing 15. https://saiia.org.za/wp-content/uploads/2010/02/Policy-Briefing-15.pdf (Accessed 19 March 2024).

Games, D. 2011. 'Doing Business in Post-Conflict and Fragile S tates: Challenges and Risks'. Development Bank of Southern Africa. https://www.dbsa.org/sites/default/files/media/documents/2022-11/Doing%20Business%20in%20Post-conflict%20and%20Fragile%20States%20-%202011.pdf (Accessed 19 March 2024).

Games, D. 2012. 'South Africa as Africa's Gateway: A Perspective from Business'. SAIIA Policy Briefing 46. https://saiia.org.za/wp-content/uploads/2012/03/Policy-Briefing-46.pdf (Accessed 19 March 2024).

Games, D. 2015. 'MTN Saga Presents Space to Rearrange SA-Nigeri a Fraught Ties'. *Business Day*, 9 November. Available at: http://www.bdlive.co.za/opinion/columnists/2015/11/09/mtn-saga-presents-space-to-rearrange-sa-nigeria-fraught-ties (Accessed 31 March 2016).

Ganywa, E., and Tshilombo, B. 1999. 'The Congo', in E. M. Noam (ed.), *Telecommunications in Africa*. Oxford: Oxford University Press on Demand, 122–9.

Gatinois, C. 2014. 'Portugal Indebted to Angola after Economic Reversal of Fortune'. *The Guardian*, 3 June. Available at: http://www.theguardian.com/world/2014/jun/03/portugal-economy-bailout-angola-invests (Accessed 1 March 2016).

Gedye, L. 2007. 'No Easy Walk to Cable Freedom'. *The M&G Online*. Available at: https://mg.co.za/article/2007-07-23-no-eassy-walk-to-cable-freedom (Accessed 15 July 2018).

Gedye, Lloyd, and Matuma Letsoala. 2009. 'Vodacom: ANC's Elephant Hunt'. *The Mail & Guardian*. 25 May. https://mg.co.za/article/2009-05-25-vodacom-ancs-elephant-hunt/ (Accessed 18 March 2024).

Geldenhuys, D. 1982. 'South Africa and the West', in R. Schrire (ed.), *South Africa: Public Policy Perspectives*. Cape Town: Juta, 306–20.

Geldenyhuys, D. 1984. *The Diplomacy of Isolation: South African Foreign Policy Making*. Johannesburg: Macmillan South Africa.

Gerring, J. 2004. 'What Is a Case Study Good For?' *American Political Science Review*, 98(2), 341–54.

Ghirwa, F. S. 2004. 'Do South African Corporations Play by the Rules?: Selected Case Studies from Zambia and Kenya'. Stability, Poverty Reduction and South African Trade and Investment in Southern Africa. Southern African Regional Poverty Network. http://www.tanzaniagateway.org/docs/Stability_Poverty_Reduction_and_SA_Trade_Investment_papers_March2004.pdf (Accessed 19 March 2024).

Gibb, R. 2009. 'Regional Integration and Africa's Development Trajectory: Meta-Theories, Expectations and Reality'. *Third World Quarterly*, 30(4), 701–21.

Gilbert, P. 2016. 'MTN Shares Fall on Nigerian Parliamentary Probe'. *IT Web*, 29 March. Available at: http://www.itweb.co.za/index.php?option=com_content&view=article&id=151019 (Accessed 31 March 2016).

Gillwald, Alison. 2002. 'Under-serviced Area Licences in South Africa: Steps to Achieving Viable Operators'. LINK Centre, University of the Witwatersrand, Johannesburg. https://www.researchgate.net/profile/Alison-Gillwald/publication/228385962_Under-serviced_Area_Licences_in_South_Africa_Steps_to_achieving_viable_operators/links/0046352c400b5af77f000000/Under-serviced-Area-Licences-in-South-Africa-Steps-to-achieving-viable-operators.pdf (Accessed 23 March 2024).

Goldstuck, A. 2006. *The Hitchhikers Guide to Going Mobile*. Cape Town: Double Storey Books.

Graham, M. 2011. 'Coming in from the Cold: The Transitional Executive Council and South Africa's Reintegration into the International Community'. *Journal of Commonwealth and Comparative Politics*, 49(3): 359–78.

Grant, C. 2011. 'State Visits as a Tool of Economic Diplomacy: Bandwagon or Business Sense'. SAIIA Occasional Paper 87. https://saiia.org.za/wp-content/uploads/2011/07/Occasional-Paper-87.pdf (Accessed 18 March 2024).

Grant, C. 1985. 'The Banks Abandon South Africa'. *Euromoney*, December, 63–77.

Grant Makokera, C. 2015. 'South African Economic Diplomacy'. ISS Paper 280. https://issafrica.s3.amazonaws.com/site/uploads/Paper280.pdf (Accessed 18 March 2024).

Great Britain Colonial Office. 1935. 'Colonial Reports – Annual'. Issues 1727–1750. London: H.M. Stationery Office.

Griffin Advisors. 2016. 'Sluggish Local Economy Impels SA Companies to Expand into the Rest of the Continent'. Available at: http://website.griffin-advisors.com/wp-content/uploads/2016/05/Griffin-Advisors-Article-21-April-2016.pdf (Accessed 20 January 2017).

Grobbelaar, N. 2004. *Every Continent Needs an America: The Experience of South African Firms Doing Business in Mozambique*. Johannesburg: South African Institute of International Affairs.

Grobbelaar, N. 2006. 'South African Corporate Engagement with Africa: Unpacking Negative and Positive Perceptions', in *South African Yearbook of International Affairs*. Johannesburg: South African Institute of International Affairs, 65–75.

Grobbelaar, N. 2014. 'Rising Powers in International Development in the state of the Debate in South Africa'. *IDS Evidence Report 91*. https://opendocs.ids.ac.uk/opendocs/bitstream/handle/20.500.12413/4305/ER91%20Rising%20Powers%20in%20International%20Development%20The%20State%20of%20the%20Debate%20in%20South%20Africa.pdf;jsessionid=7E2B76DFA011BD59613E33E1C05454F4?sequence=1 (Accessed 19 March 2024).

Gruzd, S. 2016. Interview at SAIIA Johannesburg Office. (O. P. Ayodele, Interviewer).

GSMA. 2016. 'Tech Hubs in Africa'. Available at: http://www.gsma.com/mobilefordevelopment/wp-content/uploads/2016/08/Tech-Hubs-in-Africa_Infographic.pdf (Accessed 2 February 2017).

GSMA. 2022. 'The Mobile Economy Sub-Saharan Africa 2022 – GSMA'. GSMA. Available at: https://www.gsma.com/mobileeconomy/wp-content/uploads/2022/10/The-Mobile-Economy-Sub-Saharan-Africa-2022.pdf (Accessed 29 October 2023).

GSMA Intelligence. 2014. 'The Mobile Economy: Sub-Saharan Africa'. GSMA.

GSMA Intelligence. 2015. 'The Mobile Economy: Sub-Saharan Africa'. GSMA.

Haftu, G. G. 2019. 'Information Communications Technology and Economic Growth in Sub-Saharan Africa: A Panel Data Approach'. *Telecommunications Policy*, 43(1), 88–99.

Hamill, J. 2016. 'The African Renaissance versus the South African Renaissance?' *Adelphi Series*, 56(463), 79–106.

Hamill, J., and Lee, D. 2001. 'A Middle Power Paradox? South African Diplomacy in the Post-Apartheid Era'. *International Relations*, 15(4): 35–59.

Handley, A. 2008. *Business and the State in Africa: Economic Policy-Making in the Neo-liberal Era*. New York: Cambridge University Press.

Harris, Robert G., and Kraft, C. Jeffrey. 1997. 'Meddling Through: Regulating Local Telephone Competition in the United States'. *Journal of Economic Perspectives*, 11 (4), 93–112.

HCI. (n.d.). 'Hosken Consolidated Investment Limited'. Available at: https://www.hci.co.za/ (Accessed 1 August 2023).

Headrick, D. R. 1981. *The Tools of Empire: Technology and European Imperialism in the Nineteenth Century*. New York: Oxford University Press.

Headrick, D. R., and Griset, P. 2001. 'Submarine Telegraph Cables: Business and Politics, 1838–1939'. *Business History Review*, 75(3), 200, 543–78.

Hentz, J. J. 2000. 'The Two Faces of Privatisation: Political and Economic Logics in Transitional South Africa'. *Journal of Modern African Studies*, 38(2), 203–23.

Henwood, R. 1997. 'South Africa's Foreign Policy: Principles and Problems'. Institute for Strategic Studies Monograph 13. https://issafrica.s3.amazonaws.com/site/uploads/Mono13.pdf (Accessed 18 March 2024).

Hills, J. 1990. 'The Telecommunications Rich and Poor'. *Third World Quarterly* 12(2), 71–90.

Hills, J. 1994. 'Dependency Theory and Its Relevance Today: International nstitutions in Telecommunications and Structural Power'. *Review of International Studies* 20(2), 169–86.

Hirsch, A., Levy, B., and Nxele, M. 2021. 'Politics and Economic Policymaking in South Africa Since 1994', in A. Oqubay, F. Tregenna, and I. Valodia (eds), *The Oxford Handbook of the South African Economy*. Oxford: Oxford University Press, 66–90.

Hofstatter, S., Rose, R., and wa Afrika, M. 2013. 'Report Exposes MTN "Gifts"'. *Times Live*, 15 March. Available at: http://www.timeslive.co.za/specialreports/stinvestigations/2013/03/15/report-exposes-mtn-gifts?service=print (Accessed 9 September 2015).

Hopkins, J. C., and Halstead, M. 2012. *South Africa, South Africa and the Boer-British War, Volume I Comprising a History of South Africa and Its people, Including the War of 1899 and 190*. Available at: http://www.gutenberg.org/files/41521/41521-h/41521-h.htm?fbclid=IwAR3I5blmpDnRKXsMdYAdupVIKBghMX5AisJqX84peB7HTMq0tBQBB_hmb_M (Accessed 30 July 2023).

Horwitz, R. B. 1999. 'South African Telecommunications: History and Prospects'. *Telecommunications in Africa*, 205–48.

Horwitz, R. B. 2001a. *Communication and Democratic Reform*. Cambridge: Cambridge University Press.

Horwitz, R. B. 2001b. '"Negotiated Liberalization" the Politics of Communications Sector Reform in South Africa', in N. Morris (ed.), *Media and Globalization*. New York: Rowman & Littlefield.

Horwitz, R. B., and Currie, W. 2007. 'Another Instance Where Privatisation Trumped Liberalization: The Politics of Telecommunications Reform in South Africa – A Ten Year Retrospective'. *Telecommunications Policy*, 31(8–9), 445–62.

House of Commons. 1913. 'New Marconi Agreement'. 8 August, Debate. *Hansard* 56, cc 1939–2031.

House of Commons. 1924. 'Post Office Contracts [Marconi Wireless Telegraph Company, Limited]'. 1 August Debate. *Hansard* 176, cc 2470–85.

Houston, G., Liebenberg, I., and Dichaba, W. 2001. 'Interest Group Participation in the National Economic Development and Labour Council', in G. F. Houston (ed.), *Public Participation in Democratic Governance in Southern Africa*. Pretoria: HSRC.

Hudson, J. 2007. 'South Africa's Economic Expansion into Africa: Neo-colonialism or Development', in A. Adebajo, A. Adedeji and C. Landsberg (eds), *South Africa in Africa: The Post-Apartheid Era*. Scottsville: University of KwaZulu-Natal Press.

I-Net Bridge. 2010. 'MTN Nigeria Signs $2.2bn Loan Deal'. *Fin24*, 10 June. Available at: http://www.fin24.com/Companies/MTN-Nigeria-signs-22bn-loan-deal-20100610 (Accessed 24 December 2015).

Ibukun, Y., Potelwa, X., and Onu, E. 2016. 'Nigeria Flirts with Economic Disaster as Naira Controls Stay'. *Bloomberg*, 26 January. Available at: http://www.bloomberg.com/news/articles/2016-01-26/nigeria-flirts-with-economic-disaster-as-currency-controls-stay (Accessed 1 February 2016).

ICASA (Independent Communications Authority of South Africa). 2023. 'The State of the ICT Sector Report in South Africa, ICASA'. Available at: https://www.ellipsis.co.za/wp-content/uploads/2023/07/State-of-ICT-Sector-Report-31-March-2023-updated.pdf (Accessed 1 December 2023).

ICSID. (n.d.). 'ICSID Additional Facility Rules'. Available at: https://icsid.worldbank.org/apps/ICSIDWEB/icsiddocs/Pages/ICSID-Additional-Facility-Rules.aspx (Accessed 29 October 2015).

Ifeakandu, A. 2010. 'Nigeria: MTN Signs N318 Billion Loan Facilities with Banks'. *AllAfrica*, 9 June. Available at: http://allafrica.com/stories/201006100768.htm (Accessed 24 December 2015).

Iheduru, O. C. 2008. 'Why "Anglo Licks the ANC's Boots": Globalisation and State-Capital Relations in South Africa'. *African Affairs*, 107(428): 333–60.

Ikimẹ, O. 1973. 'The British "Pacification" of the TIV 1900–1908'. *Journal of the Historical Society of Nigeria*, 7(1), 103–9.

Institute of Directors in Southern Africa. 2009. 'King III Report on Corporate Governance for South Africa Simplified Version'. https://cdn.ymaws.com/www.iodsa.co.za/resource/collection/94445006-4F18-4335-B7FB-7F5A8B23FB3F/King_III_Code_for_Governance_Principles_.pdf (Accessed 18 March 2024).

International Court of Justice (ICJ). 1962. 'Preliminary Objections', 21 December. Available at: http://www.refworld.org/cases,ICJ,402 3a86c4 html (Accessed 17 August 2018).

International Monetary Fund (IMF). 2015. *World Economic Outlook: Adjusting to Lower Commodity Prices*. Washington, DC: International Monetary Fund.

International Telecommunication Union (ITU). 1865. 'International Telegraph Conference, Paris'.

International Telecommunication Union (ITU). 1868. 'International Telegraph Conference, Vienna'.

International Telecommunication Union (ITU). 1871–2. 'International Telegraph Conference, Rome'.

International Telecommunication Union (ITU). 1875. 'International Telegraph Conference, St. Petersburg'.

International Telecommunication Union (ITU). 1932. 'International Telegraph Conference, Madrid'.

International Telecommunication Union (ITU). 1947. 'International Telecommunications Conference, Atlantic City'.

International Telecommunication Union (ITU). 1952. 'Plenipotentiary Conference, Buenos Aires'.

International Telecommunication Union (ITU). 1959. 'Plenipotentiary Conference, Geneva'.

International Telecommunication Union (ITU). 1965. 'Plenipotentiary Conference, Montreux'.

International Telecommunication Union (ITU). 1973. 'Plenipotentiary Conference, Málaga-Torremolinos'.

International Telecommunication Union (ITU). 1982. 'Plenipotentiary Conference, Nairobi'.

International Telecommunication Union (ITU). 1989. 'Plenipotentiary Conference, Nice'.

International Telecommunication Union (ITU). 1992a. 'Constitution and Convention of the International Telecommunication Union (with annexes and optional protocol)'. Concluded at Geneva on 22 December 1992.

International Telecommunication Union (ITU). 1992a. 'Additional Plenipotentiary Conference, Geneva'.

International Telecommunication Union (ITU). 1994. 'Plenipotentiary Conference, Kyoto'.

International Telecommunication Union (ITU). 1998. 'Plenipotentiary Conference, Minneapolis'.

International Telecommunication Union (ITU). 2009. 'Information Society Statistical Profiles: Africa'. Addis Ababa: International Telecommunication Union.

International Telecommunication Union (ITU). 2015a. 'ICT Facts & Figures: The World in 2015'. Available at: https://www.itu.int/en/ITU-D/Statistics/Documents/facts/ICTFactsFigures2015.pdf (Accessed 12 May 2016).

International Telecommunication Union (ITU). 2015b. 'Status of the Transition to Digital Terrestrial Television Broadcasting'. International Telecommunications Union, 20 August. Available at: http://www.itu.int/en/ITU-D/Spectrum-Broadcasting/Pages/DSO/Default.aspx (Accessed 20 August 2015).

International Telecommunications Union (ITU). 2016. 'ICT Facts & Figures: The World in 2016'. Available at: http://www.itu.int/en/ITU-D/Statistics/Documents/facts/ICTFactsFigures2016.pdf (Accessed 12 February 2017).

Intven, H., and McCarthy, T. 2005. *Universal Access: How to Reach the Last 20 Percent: A Primer*. Washington, DC: World Bank. Available at: https://www.itu.int/ITU-D/treg/Documentation/Infodev_handbook/6_Univ_Service.pdf (Accessed 27 July 2023).

IOL. 2012. 'Cosatu Calls for a Sea-Change on Graft'. *Independent Newspapers Online*, 24 June. Available at: http://www.iol.co.za/news/politics/cosatu-calls-for-a-sea-change-on-graft-1.1326120#.TcIbRfrr64 (Accessed 24 June 2012).

IRIN. 2009. 'Growth Slips over Oil Price'. *IRIN News*, 29 January. Available at: http://www.irinnews.org/news/2009/01/29/growth-slips-over-oil-price (Accessed 1 March 2016).

Isa, M. 2002. 'SA Launches Good News Assault on Sick Rand'. *Independent Newspapers Online*, 8 January. Available at: http://www.iol.co.za/news/south-africa/sa-launches-good-news-assault-on-sick-rand-80329 (Accessed 3 June 2016).

IT News Africa Staff. 2008. 'President Motlanthe Signs TIA Bill to Boost Telecommunications Innovation in SA'. *IT News Africa*, 28 November. Available at: http://www.itnewsafrica.com/2008/11/president-motlanthe-signs-tia-bill-to-boost-telecommunications-innovation-in-sa/ (Accessed 31 March 2016).

IT News Africa Staff. 2014. 'Turkcell Once Again Loses Court Challenge against MTN'. *IT News Africa*, 20 October. Available at: http://www.itnewsafrica.

com/2014/10/turkcell-once-again-dominated-by-mtn-group/ (Accessed 31 March 2016).

Jacobs, A. 1948. *South African Heritage: A Biography of H.J. Van der Bijl*. Pietermaritzburg: Shuter & Shooter Publishers.

Jacobs, S., and Duarte, D. 2010. 'Protest in Mozambique: The Power of SMS'. *Afronline*, 16 September. Available at: http://www.afronline.org/?p=8680 (Accessed 1 January 2016).

Joffee, H. 2015. 'What's Hampering State-Business Co-operation. Trust Deficit or Lack of Will'. *Rand Daily Mail*, 9 September. Available at: http://www.rdm.co.za/business/2015/09/09/what-s-hampering-state-business-coperation-tust-deficit-or lack-of-will (Accessed 9 September 2015).

Jones, G. 2013. 'SA Companies Spend More on Social Investment'. *Business Day BDlive*, 4 December. Available at: http://www.bdlive.co.za/business/2013/12/04/sa-companies-spend-more-on-social-investment (Accessed 10 October 2015).

Jurgens, A. 2011. 'Research Report on Telecommunications: Who Owns Whom'. https://www.whoownswhom.co.za/products/woweb-platform/ (Accessed 19 March 2024).

Kahn, B. 1991. The Crisis and South Africa's Balance of Payments', in Stephen Gelb (ed.), *South Africa's Economic Crisis*. Cape Town: David Philip, 1–32.

Kapferer, B. 2004. Introduction: Old permutations, New Formations? War, State, and Global Transgression. *Social Analysis*, 48(1), 64–96.

Kazeem, Y. 2020. 'African Mobile Users Pay Nearly Three Times the Global Average for Voice Calls and Internet'. *Quartz*. Available at: https://qz.com/africa/1878749/how-much-do-africans-pay-for-voice-calls (Accessed 30 October 2023).

Keohane, R. O. 2009. 'The olde and IPE and the new'. *Review of International Political Economy*, 16(1): 34–46.

Keohane, R. O., and Nye, J. S. 1987. 'Power and Interdependence Revisited'. *International Organization*, 41(4).

Keohane, R. O., and Nye, J. S. 1998. 'Power and Interdependence in the Information Age'. *Foreign Affairs*, 77(5): 81–94.

Kganyago, L. 2008. 'South African Corporate Expansion into Africa: A View from the South African Government', in N. Grobblaar and H. Besada (eds), *Unlocking Africa's Potential: The Role of Corporate South Africa in Strengthening Africa's Private Sector*. Johannesburg: South African Institute of International Affairs.

Khadiagala, G. M. 2001. 'Foreign Policy Decision-Making in Southern Africa's Fading Frontline', in G. M. Khadiagala and T. Lyons (eds), *African Foreign Policies: Power and Process*. Boulder, CO: Lynne Rienner.

Khan, Firoz, and Mohamed, Seeraj. 2023 'Elites and Economic Policy in South Africa's Transition and Beyond', *International Review of Applied Economics*, 37(6), 1–23.

Kim, I. S., and Milner, H. V. 2019. 'Multinational Corporations and Their Influence through Lobbying on Foreign Policy'. *Multinational Corporations in a Changing Global Economy*, 497–536. https://www.brookings.edu/wp-content/uploads/2019/12/Kim_Milner_manuscript.pdf (Accessed 18 March 2024).

King Commission. 2009. 'King III Report on Corporate Governance for South Africa'. https://www.pwc.co.za/en/assets/pdf/king-iv-steering-point.pdf (Accessed 18 December 2015).

King Commission. 2016. 'King IV Report on Corporate Governance for South Africa'. https://www.pwc.co.za/en/assets/pdf/king-iv-steering-point.pdf (Accessed 22 December 2020).

Knott-Craig, A. 2012. 'Reflections on the Cellular Telephone Industry in South Africa by a Key Role Player'. *Focus*, 66. https://hsf.org.za/publications/focus/focus-66/AKnott-Craig.pdf (Accessed 18 December 2015).

Kohli, T. 2015. 'Why Kenya, Home to Africa's "Silicon Valley", Is Set to Be the Continent's Ultimate Tech Hub'. *Mail & Guardian Africa*, 19 February. Available at: http://mgafrica.com/article/2015-02-19-why-kenya-is-africas-tech-hub (Accessed 1 March 2016).

Kone, H. 1999. 'The Ivory Coast (Côte d'Ivoire)', in E. M. Noam (ed.), *Telecommunications in Africa*. Oxford: Oxford University Press on Demand, 141–62.

Korhonen, M. 2018. 'Factsheet: Measuring South Africa's (Black) Middle Class'. *Africa Check*. Available at: https://africacheck.org/fact-checks/factsheets/factsheet-measuring-south-africas-black-middle-class (Accessed 30 October 2023).

Kornegay, F., and Landsberg, C. 1998. 'Phaphama iAfrika!: The African Renaissance and Corporate South Africa'. *The African Security Review*, 7(4).

Kotch, N. 2014. 'Kagame, Zuma Try to End Diplomatic Dispute'. *Business Day BDlive*, 27 March. Available at: http://www.bdlive.co.za/africa/africannews/2014/03/27/kagame-zuma-try-to-end-diplomatic-dispute (Accessed 20 September 2015).

Krapohl, S., and Fink, S. 2013. 'Different Paths of Regional Integration: Trade Networks and Regional Institution-Building in Europe, Southeast Asia and South Africa'. *Journal of Common Market Studies*, 51(3), 472–88.

Kutz, M. R. 2008. 'Contextual Intelligence: An Emerging Competency for Global Leaders'. *Regent Global Business Review*, 2(2).

Lancaster, H. 2022. *Mobile Network Operators and MVNOs (Africa)* (thirteenth edn). Paul Budde Communication. https://www.budde.com.au/Research/Africa-Mobile-Network-Operators-and-MVNOs.

Landsberg, C. 2004. *The Quiet Diplomacy of Liberation: International Politics and South Africa's Transition*. Johannesburg: Jacana Media.

Landsberg, C. 2006. 'New Powers for Global Change? South Africa's Global Strategy and Status'. *FES Briefing Paper*, (16).

Landsberg, C. 2007. 'The New Afrabia: South Africa's African Agenda and the Globalisation of the South', in A. Adebajo, A. Adedeji and C. Landsberg (eds), *South Africa in Africa: The Post-Apartheid Era*. Scottsville: University of KwaZulu-Natal Press.

Langalanga, A. 2014. 'The Reed That Lies Low in a Flooded River: The Case of Swazi Resistance to South African Reform Diplomacy'. Policy Briefing 89.

Laurence, P. 2018. 'Botha Refuses to Reform Apartheid Policy – Archive 16 August 1985'. *The Guardian*. Available at: https://www.theguardian.com/world/2018/aug/16/defiant-botha-refuses-to-pledge-reform-of-apartheid-policy-archive-1985 (Accessed 25 July 2023).

Le Pere, G. 2013. 'South African Foreign Policy in a Disorderly World: Will the Centre Hold?'. *The Thinker*, 53, 20–2.

Le Pere, G. 2017. 'Ubuntu as Foreign Policy: The Ambiguities of South Africa's Global Identity'. *The Thinker*, 71, 16–19.

Levy, B., Hirsch, A., Naidoo, V., and Nxele, M. 2021. *South Africa: When Strong Institutions and Massive Inequalities Collide* (Endowment for International Peace). Cape Town: Carnegie.

Lewis, C. 2013. 'Universal Access and Service Interventions in South Africa: Best Practice, Poor Impact'. *African Journal of Information and Communication*, (13), 15–34. Available at: https://ajic.wits.ac.za/article/download/13679/18282/80808 (Accessed 27 July 2023).

Lewis, C. 2018. 'The digital Donga: Universal Access and Service in South Africa (1994–2014)'. University of the Witwatersrand. (Doctoral dissertation).

Lewis, P. 2001. 'Leon Sullivan, 78, Dies; Fought Apartheid'. *New York Times*, 26 April.

Liu, Z. 2015. 'Do We Need Local Ownership Requirement for Foreign Direct Investment?' Evidence from Chinese Firms. Available at: http://www.etsg.org/ETSG2015/Papers/090.pdf (Accessed 18 December 2016).

Lourie, G. 2015. 'Ramaphosa-MTN Tax-Allegations Raise Eyebrows?'. *Techfinancials*, 12 October. Available at: http://techfinancials.co.za/2015/10/ramaphosa-mtn-tax-allegations-raise-eyebrows/ (Accessed 2 June 2016).

Louw-Vaudran, L. 2016. *Superpower Or Neocolonialist?: South Africa in Africa*. Johannesburg: Tafelberg.

Lu, Jia, and Weber, Ian. 2007. 'State, Power and Mobile Communication: A Case Study of China'. *New Media & Society*, 9, 925–44.

Lugon-Moulin, A. n.d. 'Understanding State Capture'. Available at: http://f3magazine.unicri.it/?p=402 (Accessed 1 February 2017).

Luiz, J. M., and Stephan, H. 2012. 'The Multinationalisation of South African Telecommunications Firms into Africa'. *Telecommunications Policy*, 36(8), 621–35.

Lukhele, S. 2012. 'Plum Job at MTN for Swazi King's Daughter'. *Business Report*, 9 September. Available at: http://www.iol.co.za/business/companies/plum-job-at-mtn-for-swazi-king-s-daughter-1.1378665#.VgPAybRo2fQ (Accessed 9 September 2015).

Lyall, F. 1994. 'The International Telecommunication Union and Development'. *Journal of Space Law*, 22(23).

Lyall, F. 2011. *International Communications: The International Telecommunication Union and the Universal Postal Union*. London: Routledge.

M-Cell. 2001. 'Annual Report'.

M-Cell. 2002. 'Annual Report'.

M-Cell/MTN Group Corporate Affairs. 2002. 'Sustainable Development Report'.

Maake, M. 2015. 'Stanbic IBTC gets $5m fine'. *Business Day*, 2 November. Available at: http://www.bdlive.co.za/business/financial/2015/11/02/stanbic-gets-court-relief-in-nigeria (Accessed 31 March 2016).

Maake, M. 2016. 'Stanbic Gets Court Relief in Nigeria'. *Business Day BDLive*, 28 January. Available at: http://www.bdlive.co.za/business/financial/2016/01/28/stanbic-gets-court-relief-in-nigeria (Accessed 3 June 2016).

Maasdorp, G. 1982. 'Industrial Decentralisation and the Economic Development of the Homelands', in R. Schrire (ed.), *South Africa: Public Policy Perspectives*. Cape Town: Juta.

MacLean, D. 2003. 'The Quest for Inclusive Governance of Global ICTs: Lessons from the ITU in the Limits of National Sovereignty'. *Information Technologies & International Development*, 1(1), 1–18.

Mafeje, A. 1988. 'South Africa: The Dynamics of a Beleagured State', in B. Magubane and I. Mandaza (eds), *Whither South Africa?* Trenton: Africa World Press.

Mahlaka, R. 2015. 'Global Tech Companies Eye Stellenbosch Talent'. *Moneyweb*. Available at: https://www.moneyweb.co.za/news/tech/global-tech-companies-eye-stellenbosch-talent/ (Accessed 1 February 2017).

Maketa, T. L. 2009. 'MTN Entry Strategy in Nigeria: Case Study Research Report'. University of Cape Town, MBA Dissertation, The Graduate School of Business. UCT.

Makgoba, M. W. 2020. 'The African Renaissance Conference, Twenty-One Years Later: A Critical Reflection'. *International Journal of African Renaissance Studies – Multi-, Inter- and Transdisciplinarity*, 15(1), 59–72.

Manco Voice. 2011. 'MTN Shares Latest Trends with SA Ambassadors Based at Our Opcos'. MTN Group Website, July. Available at: https://www.mtn.com/Documents/NewsLetter/share-latest-trends.html?utm_source=MancoNewsletter&utm_medium=Email&utm_content=July2011Issue3&utm_campaign=MTN_MancoEmailNewsletter (Accessed 5 June 2016).

Mandela, N. 1995. 'Address by President Nelson Mandela at the Opening Ceremony of Telecom 95, the 7th World Telecommunications Forum and Exhibition, Geneva'. ANC, 3 October. Available at: http://www.anc.org.za/show.php?id=3608 (Accessed 31 May 2016).

Mandela, N. 1995. 'Address by President Nelson Mandela at the Opening of Telecom 95, the 7th World Telecommunications Forum and Exhibition'. Available at: http://www.mandela.gov.za/mandela_speeches/1995/951003_telecom.htm (Accessed 15 July 2018).

Mangome, M. 2012. 'State Can't Afford Wage Hikes'. *Independent Newspapers Online*, 20 June. Available at: http://www.iol.co.za/pretoria-news/state-can-t-afford-wage-hikes-sisulu-1.1323198#.T-cH5Bfrr64 (Accessed 20 June 2012).

Mantshanthsa, S. 2015. 'MTN to Spend R30bm on Infrastructure Upgrades'. *Business Day Live*, 5 March. Available at: http://www.bdlive.co.za/business/technology/2015/03/05/mtn-to-spend-r30bn-on-infrastructure-upgrades (Accessed 2 August 2015).

Marais, J., and Rose, R. 2013. 'Iran Deals Forced MTN Boss to Quit'. *Business Day BDlive*, 28 July. Available at: http://www.bdlive.co.za/business/technology/2013/07/28/iran-deals-forced-mtn-boss-to-quit?service=print (Accessed 1 August 2015).

Maree, J. 1993. 'Trade Unions and Corporatism in South Africa'. *Transformation*, 21. https://transformationjournal.org.za/wp-content/uploads/2017/03/tran021003.pdf (Accessed 18 December 2015).

Maroleng, C. 2016. Interview Held at MTN Innovation Centre, Fairlands. (O. P. Ayodele, Interviewer).

Marthoz, J.-P. 2012. 'The Challenges and Ambiguities of South Africa's Foreign Policy'. The Norwegian Peace Building Resource Centre. Available at: https://www.files.ethz.ch/isn/153069/1d25c90556f0a6f66548551220c882e8.pdf (Accessed 10 February 2017).

Masango, S. 2015. *Towards a Participatory Foreign Policy*. Braamfontein: SARChI Chair in African Diplomacy, University of Johannesburg and the Friedrich-Ebert-Stiftung South Africa.

Masmoudi, M. 1978. 'The New World Information Order: Document'. Working Paper 31. Paris: UNESCO.

Masters, L. 2012. 'Opening the "Black Box": South African Foreign Policy-Making', in C. Landsberg and J.-A. van Wyk (eds), *South African Foreign Policy Review* (vol. 1). Pretoria: Africa Institute of South Africa.

Masters, L. 2016. Interview Held at the University of Johannesburg. (O. P. Ayodele, Interviewer).

Matlala, G., Hawker, D., and Mofokeng, M. 2012. 'ANC Set to Scrap E-Tolls'. *Independent Newspapers Online*. Available at: http://www.iol.co.za/motoring/industry0news/anc-set-to-scrap-et-tolls-1.12.85710#T-cIORfrr64 (Accessed 25 June 2012).

Matlosa, K. 2007. 'South Africa and Regional Security in Southern Africa', in A. Adebajo, A. Adedeji and C. Landsberg (eds), *South Africa in Africa: The Post-Apartheid Era*. Scottsville: University of KwaZulu-Natal Press.

Mawson, N. 2013. 'Ramaphosa Steps Down as MTN Chairman'. *ITWeb*, 6 March. Available at: http://www.itweb.co.za/index.php?option=com_content&view=article&id=62269 (Accessed 1 August 2015).

Mays, T. M. 2015. *Historical Dictionary of International Organizations in Africa and the Middle East*. Lanham: Rowman & Littlefield.

Mbeki, T. 1995. 'South Africa and the Information Superhighway, Keynote Address to the G7 Ministerial Conference on the Global Information Society'. Brussels, Belgium, 24 February 1995. Available at: http://aei.pitt.edu/33414/1/A161.pdf (Accessed 14 July 2018).

Mbeki, T. 1996. 'Statement of Deputy President Thabo Mbeki, on Behalf of the African National Congress, on the Occasion of the Adoption by the Constitutional Assembly of "The Republic of South Africa Constitution Bill 1996" '. https://web.archive.org/web/20060615053925/http://www.anc.org.za/ancdocs/history/mbeki/1996/sp960508.html (Accessed 14 June 2012).

Mbeki, T. 1998. 'The African Renaissance Statement of Deputy President, Thabo Mbeki'. The African Renaissance Statement of deputy president Thabo Mbeki, SABC Gallagher Estate. https://web.archive.org/web/20021020041906/http://www.anc.org.za/ancdocs/history/ mbeki/1998/tm0813.htm (Accessed 18 December 2015).

Mbungela, M. 2016. 'Vodacom Embraces Change, Latest Stories'. Available at: https://now.vodacom.co.za/article/vodacom-embraces-change (Accessed 4 August 2023).

McCabe, D. 2016. 'Bill Clinton's Telecom Law: Twenty Years Later'. *The Hill*. https://thehill.com/policy/technology/268459-bill-clintons-telecom-law-twenty-years-later/ (Accessed 20 December 2020).

McCarthy, C. 2008. 'The Challenge of Reconciling Revenue Distribution and Industrial Development in South Africa', in B. Vickers (ed.), *Industrial Policy in the Southern African Customs Union: Past Experiences, Future Plans*. Midrand: Institute for Global Dialogue.

McKune, C., and Turner, G. 2015. 'MTN Bullies SA with Jobs Threats'. *Mail & Guardian*, 16–22 October.

McLeod, D. 2015. 'Cwele Must Seize Rare Opportunity'. *TechCentral*, 9 August. Available at: http://www.techcentral.co.za/time-for-wholesale-change-at-dtps/58834/ (Accessed 9 August 2015).

Meineke, M. 2013. 'RLPC-Banks See Big Opportunities in Nigerian Syndicated Loans'. *Reuters*, 29 April. Available at: http://www.reuters.com/article/nigeria-loans-idUSL6N0DG28220130429 (Accessed 26 December 2015).

Melber, H. 2004. 'South Africa and NEPAD – Quo Vadis'. *Policy Brief*, 31, 1–11.

Melber, H. 2017. 'The African Middle Class(es) – in the Middle of What?'. *Review of African Political Economy*, 44(151), 142–54.

Michaelis, A. R. 1965. *From Semaphore to Satellite*. Geneva: International Telecommunication Union.

Miller, D., Saunders, R., and Oloyede, O. 2008. 'South African Corporations and Post-Apartheid Expansion in Africa: Creating a New Regional Space'. *African Sociological Review*, 12(1), 1–19.

Miller, J. 2009. 'Soft-Power and State-Firm Diplomacy: Congress and IT Corporate Activity in China'. *International Studies Perspectives*, 10(3), 285–302.

Mills, G. 1997. 'India & South Africa: The Search for Partnership'. *SAIIA Report*.

Mills, G. 2002. 'Prisoners of a Paradigm? Understanding SA's Foreign Policy'. South African Yearbook of International Affairs 2002–3.

Minges, M., and Kelly, T. 1994. 'The Paradoxes of African Telecommunications', in B. A. Kiplagat and M. C. M. Werner (eds), *Telecommunications and Development in Africa*, 11–29. Oxford: IOS Press.

Miroux, A. 2008. 'Transnational Corporations and the Infrastructure Challenge'. NEPAD-OECD Africa Investment Initiative 2008 High-Level Meeting /Expert Roundtable, Kampala, Uganda, 10–11 December 2008. https://www.oecd.org/investment/investmentfordevelopment/41881230.pdf.

Miti, K., and Kilambo, S. R. 2012. 'South Africa as a Gateway to Africa: Dream or Reality'. *Insight on Africa*, 4(1), 59–67.

Mlumbi-Peter, X. 2016. 'Concluding Remarks by Ms Xolelwa Mlumbi-Peter'. Pretoria: Department of Trade and Industry. Available at: http://www.thedti.gov.za/trade-invest/DDG_Closing_Remarks.pdf (Accessed 3 August 2016).

Mochiko, T. 2015. 'PIC Turns Up Heat as Nhleko in MTN Hot Spot'. *Business Day*.

Mochiko, T. 2017. 'Cell C: Bad Blood Over BEE'. *BusinessLIVE*. Available at: https://www.businesslive.co.za/fm/features/2017-07-06-cell-c-bad-blood-over-bee/ (Accessed 28 July 2023).

Mokgata, Z. 2008. 'Vodacom Goes Ahead with Empowerment Deal'. *SowetanLIVE*. Available at: https://www.sowetanlive.co.za/news/2008-04-01-vodacom-goes-ahead-with-empowerment-deal/ (Accessed 1 August 2023).

Moons, S., and van Bergeijk, P. 2013. 'A Meta-Analysis of Economic Diplomacy and Its Effect on International Economic Flows'. ISS Working Paper Series/General Series, 566, 1–30.

Moshoshoe, M. 1991. 'Africa in a Post-Apartheid Era'. The Challenges of Post-Apartheid South Africa: The African Leadership Forum. Windhoek: African Leadership Forum.

Mothlante, K. 2008. 'Input Made by President Kgalema Motlanthe at the G-20 Summit on Financial Markets and the World Economy'. ANC. Available at: http://www.anc.org.za/show.php?id=4032 (Accessed 3 June 2016).

Mpofu, F. Y. 2022. 'Taxing the Digital Economy through Consumption Taxes (VAT) in African Countries: Possibilities, Constraints and Implications'. *International Journal of Financial Studies*, 10(3), 1–21.

Mququ, S. 2016. 'Brand South Africa's Africa Strategy Development'. Johannesburg, South Africa: University of Johannesburg.

MTN. 2020. 'History'. MTN Group. 30 March. https://group.mtn.com/who-we-are/history/.

MTN. 2023. 'MTN Group FY 22 Results Data Sheets, MTN Annual Results FY22'. https://www.mtn.com/wp-content/uploads/2023/03/MTN-Group-FY-22-data-sheets.xls (Accessed 4 December 2023).

MTN Group. 2005. 'MTN Group SENS Announcement: Announcement to MTN Shareholders'. *Imara*, 6 April. Available at: http://trade.imara.co/free/sens/disp_news.phtml?tdate=20050406080001&seq=165&scheme=imaraco (Accessed 20 March 2016).

MTN Group. 2009. 'Annual Report'. https://www.mtn-investor.com/mtn_ar09/book1/pdf/full.pdf (Accessed 18 March 2024).

MTN Group. 2014. Limited Corporate Governance Report Ending 31 December.

MTN Group. 2016. 'MTN Group SENS Announcement: Disclosure of Beneficial Interests in Securities in MTN Group'. *Sharenet*, 1 February. Available at: http://

www.sharenet.co.za/jsepdf/SENS_20160201_S367711.pdf (Accessed 20 March 2016).

MTN Group Limited. 2003a. 'Annual Report'.

MTN Group Limited. 2003b. 'Sustainability Report'.

MTN Group Limited. 2004. 'Annual Report'.

MTN Group Limited. 2006. 'Annual Report for the Financial Year Ended 31 December 2006'.

MTN Group Limited. 2013a. 'Integrated Report for the Year Ended 31 December 2013'.

MTN Group Limited. 2015. 'Annual Sustainability Report for the Year Ended 31 December 2015'.

MTN Group Limited. 2022. 'Integrated Report for the Year Ended 31 December 2022'.

Muller, R. 2007a. 'Mbeki Brokers $60m Benin Deal for MTN'. *MyBroadband*, 18 September. Available at: http://mybroadband.co.za/news/cellular/1347-mbeki-brokers-60m-benin-deal-for-mtn.html (Accessed 1 August 2015).

Muller, R. 2007b. 'Nepad, Eassy Partnership Ends in Divorce'. *Mybroadband.Co.Za*. Available at: https://mybroadband.co.za/news/telecoms/827-nepad-eassy-partnership-ends-in-divorce.html (Accessed 15 July 2018).

Muller, R. 2014. 'How the ANC Nearly Killed Vodacom, MTN'. *MyBroadband*. Available at: http://mybroadband.co.za/news/cellular/98658-how-the-anc-nearly-killed-vodacom-mtn.html (Accessed 2 August 2015).

Mungadze, S. 2014. 'Uganda Says MTN Is Withholding Taxes'. *Business Day BDLive*. Available at: http://www.bdlive.co.za/africa/africanbusiness/2014/08/13/uganda-says-mtn-is-withholding-taxes (Accessed 20 September 2015).

Murphy, J. T., and Carmody, P. R. 2015. *Africa's Information Revolution: Technical Regimes and Production Networks in South Africa and Tanzania*. Chichester: Wiley-Blackwell.

Muthien, Y., and Khoza, M. 2002. 'The ICT Sector Meeting the NEPAD Challenge: M-Cell/MTN's Contribution to the African Renaissance'. *Journal of Public Administration*, 37(3.1).

MyBroadband. 2015. 'Government Eyes Vodacom Sale, MTN Mulls Telkom Buy: Report'. Available at: http://businesstech.co.za/news/telecommunications/78823/government-eyes-vodacom-sale-mtn-mulls-telkom-buy-report/ (Accessed 1 August 2015).

MyBroadband staff. 2011. 'MTN Boosts BBBEE rating'. Available at: http://mybroadband.co.za/news/business/25301-mtn-boosts-bbbee-rating.html (Accessed 22 March 2016).

Naidoo, S. 2012. 'MTN Rocked by Iran Licence Bribary Scandal'. *Mail and Guardian Online*. Available at: http://www.mg.co.za/article/2012-03-29-turkcell-sues-mtn-over-iran-contracts (Accessed 12 March 2012).

Naidu, S. 2016. 'Skype Interview'. University of Johannesburg. (O. P. Ayodele, Interviewer).

Naidu, S., and Lutchman, J. 29–30 March 2004. 'Understanding South Africa's Engagement in the Region: Has the Leopard Changed Its Spots?'. Stability, Poverty Reduction and South African Trade and Investment in Southern Africa. Southern African Regional Poverty Network. http://www.tanzaniagateway.org/docs/Stability_Poverty_Reduction_and_SA_Trade_Investment_papers_March2004.pdf.

Naidu, S., and Vickers, B. 2014. 'South Africa, Africa, and International Investment Agreements'. Stellenbosch: Centre for Conflict Resolution.

National Treasury. 2009. 'Financial Regulatory Approvals Sought by MTN SA for the Proposed Merger with Bharti Airtel, India'. http://www.treasury.gov.za/comm_media/press/2009/2009093001.pdf (Accessed 4 June 2016).

National Treasury. 2012. 'Budget Review'. Republic of South Africa. https://assets.publishing.service.gov.uk/media/57a08b2aed915d622c000b4d/dp34a.pdf.

Nattrass, N., and Seekings, J. 2010. 'State, Business and Growth in Post-Apartheid South Africa: Discussion Paper Series'. Available at: http://www.ippg.org.uk (Accessed 11 November 2011).

Nau, H. R. 2014. *Perspectives on International Relations: Power, Institutions, and Ideas*. Thousand Oaks, CA: CPQ Press.

NCC. 2019. MTN Has Paid N275bn SIM Infraction Fine, Nigerian Communications Commission News Release'. https://ncc.gov.ng/media-centre/news-headlines/553-press-release-ncc-mtn-has-paid-n275bn-sim-infraction-fine (Accessed 18 March 2024).

Ndajiwo, M. 2020, June. 'The Taxation of the Digitalised Economy: An African Study'. Working Paper 10. https://opendocs.ids.ac.uk/opendocs/bitstream/handle/20.500.12413/15427/ICTD_WP107.pdf?sequence=1 (Accessed 11 July 2020).

Ndemo, B., and Mureithi, M. 2015. 'Information and Communications Technologies', in T. McNamee, M. Pearson, and W. Boer (eds), *Africans Investing in Africa: Understanding Business and Trade, Sector by Sector*. Basingstoke: Palgrave Macmillan.

Ndung'u, M. N., and Waema, T. M. 2012. 'Understanding What Is Happening in ICT in Kenya: A Supply and Demand Side Analysis of the ICT Sector'. Policy Paper,

Research ICT Africa, Cape Town. https://researchictafrica.net/publication/understanding-what-is-happening-in-ict-in-kenya/.

NEDLAC. n.d. Available at: http://new.nedlac.org.za (Accessed 1 October 2015).

Nel, C., and Grealy, R. 1989. 'The Consultative Business Movement'. *Indicator SA*, 6(2), 105–8.

Nene, N. M. 2015. 'Medium Term Budget Policy Statement Speech'. Pretoria: Department of National Treasury.

NEPAD. 2001. 'The New Partnership for Africa's Development'. Midrand, South Africa: NEPAD Secretariat.

NEPAD e-Africa Commission. 2005. 'The Nepad ICT Infrastructure Programme Brochure'. Available at: https://event-africa-networking.web.cern.ch/event-africa-networking/cdrom/NEPAD/NEPAD_ICT_Infrastructure_Brochure_Sept05.pdf (Accessed 15 July 2018).

Netshitenzhe, J. 1998. 28 September. 'The Media in the African Renaissance'. African Renaissance Conference Session 6. Available at: https://www.gcis.gov.za/content/newsroom/speeches/ceo/joel-netshitenzhe-african-renaissance-conference-session-6 (15 July 2018).

Nganje, F. 2012. 'The influence of the ANC on South Africa's Foreign Policy'. Institute for Global Dialogue.

Nganje, F., and Letshele, L. 2020. 'Challenges and Opportunities of Non-traditional Diplomacy', in D. D. Bradlow and E. Sidiropoulos (eds), *Values, Interests and Power: South African Foreign Policy in Uncertain Times*. Pretoria University Law Press.

Ngcobo, Z. 2016. 'Sun International Employees Released from Nigeria without Charge'. *EWN*, 26 January. http://ewn.co.za/2016/01/26/Sun-international-continues-efforts-to-free-employees-detained-in-Nigeria (Accessed 18 January 2022).

Ngoma, N. 2005. *Prospects for a Security Community in Southern Africa: An Analysis of Regional Security in the Southern African Development Community.* Pretoria: Institute for Security Studies.

Nkabinde, T. 2022. 'Vodacom Achieves the Level-1 B-BBEE Status'. Available at: https://www.businesstechafrica.co.za/telecoms/2022/06/08/vodacom-achieves-the-level-1-b-bbee-status/ (Accessed 4 August 2023).

No Author. 1899. 'Cape to Cairo'. *The Brisbane Courier (Qld.: 1864–1933)*, 14 March, 5. Available at: http://nla.gov.au/nla.news-article3689941 (Accessed 22 July 2023).

Nolutshungu, S. C. 1988. 'The South African State and Africa', in B. Magubane and I. Mandaza (eds), *Whither South Africa?* Trenton: Africa World Press.

Non-Aligned Countries (NAC). 1973. '4th Summit Conference of the Non-Aligned Countries, Algiers', September, NAC/ALG/Conf.4, http://cns.miis.edu/nam/

documents/Official_Document/4th_Summit_FD_Algiers_Declaration_1973_Whole.pdf (Accessed 15 July 2018).

Ntomba, R. 2015. 'Why Are Military Coups Going Out of Fashion in Africa?'. *New African Magazine*. Available at: http://newafricanmagazine.com/why-are-military-coups-going-out-of-fashion-in-africa#sthash.jizfYhUu.dpuf (Accessed 4 February 2017).

Nurudeen, N. A. 2015. 'Many People Refused Offers to Invest in MTN Nigeria-Pascal Dozie'. *Daily Trust Online*. Available at: http://www.dailytrust.com.ng/news/business/many-people-refused-offers-to-invest-in-mtn-nigeria-pascal-dozie/108473.html (Accessed 10 September 2015).

Nyabola, N. 2018. *Digital Democracy, Analogue Politics: How the Internet Era Is Transforming Politics in Kenya*. London: Bloomsbury Publishing.

Nyambura-Mwaura, H. 2012. 'S.Africa's Shanduka Takes $335 mln Stake in MTN Nigeria'. *Reuters*. Available at: http://www.reuters.com/article/2012/11/27/shanduka-mtn-idUSL5E8MR8UV20121127 (Accessed 1 September 2015).

Nyanda, S. 2009. 'Parliamentary Question N1025'. ANC. Available at: https://www.gov.za/news/speeches/reply-minister-minister-communications-s-nyanda-questions-posed-national-assembly-2 (Accessed 8 September 2009).

O'Riordan, A. 2015. 'Is South Africa the World's Most Generous Donor?'. South African Civil Society Information Service, 18 March. Available at: http://www.sacsis.org.za/site/article/2318 (Accessed 31 May 2016).

Obasanjo, O. 1991. 'Chairman's Summary: Conference on the Challenge of Post Apartheid South Africa to Southern Africa in Particular and Africa in General', in *The Challenges of Post-Apartheid South Africa: Conclusions and Papers Presented at a Conference of the African Leadership Forum*. Windhoek: Africa Leadership Forum.

Odendaal, N. 2014. 'TIA Seeks New Leaders as CEO and CFO Dismissed'. *Engineering News*, 1 April. Available at: http://www.engineeringnews.co.za/article/tia-seeks-new-leaders-as-ceo-and-cfo-dismissed-2014-04-01 (Accessed 12 January 2017).

Odhiambo, N. M. 2022. 'Information Technology, Income Inequality and Economic Growth in Sub-Saharan African Countries'. *Telecommunications Policy* 46 (6): 1–12.

OECD, African Development Bank, United Nations Economic Commission for Africa, United Nations Development Programme. 2011. 'African Economic Outlook 2011: Africa & Its Emerging Partners'. New York: OECD Publishers.

Oguttu, A. W. 2011. 'Developing South Africa as a Gateway for Foreign Investment in Africa: A Critique of South Africa's Headquarter Company Regime'. *South African Yearbook of International Law*, 36(1), 61–93.

Ogutu, J. 2015. 'How ICT Drives Kenya's Economic Growth'. *Standard Media*, 18 May. Available at: https://www.standardmedia.co.ke/article/2000162611/how-ict-drives-kenya-s-economic-growth (Accessed 1 January 2017).

Okoroafor, C. 2015. 'Could Nigeria Be the New Locations for MTN Headquarters'. *Ventures*, 15 September. Available at: http://venturesafrica.com/could-nigeria-be-the-new-location-for-mtn-headquarters/ (Accessed 16 September 2015).

Olivier, G. 2003. 'Is Thabo Mbeki Africa's Saviour?' *International Affairs* 79(4), 815–28.

Olivier, G. C. 1982. 'South Africa's Relations with Africa', in R. Schrire (ed.), *South Africa: Public Policy Perspectives*. Cape Town: Juta.

Olivier, G., and Geldenhuys, D. 1997. 'South Africa's Foreign Policy: From Idealism to Pragmatism'. *Business & the Contemporary World*, 9(2), 363–76.

Oluwole, V. 2023. 'The High Cost of Mobile Data in Africa: Top 20 Most Expensive Countries'. *Business Insider*. Available at: https://africa.businessinsider.com/local/lifestyle/20-african-countries-with-the-most-expensive-mobile-data-prices-is-your-country-on/sjxr35p (Accessed 30 October 2023).

'The O'Malley Archives'. 2024. The Reconstruction and Development Programme (RDP) – The O'Malley Archives. https://omalley.nelsonmandela.org/index.php/site/q/03lv02039/04lv02103/05lv02120/06lv02126.htm (Accessed 17 March 2024).

Onuoha, J. C. 2008. *Economics of Knowledge Management in the Information Age: Global Challenges and Enhancement Strategies* (ed. B. G. Nworgu). Nsukka: University Trust Publishers.

Onuoha, R., and Gillwald, A. 2022. *Digital Taxation: Can It Contribute to More Just Resource Mobilisation in Post-Pandemic Reconstruction?*. Digital New Deal for Africa Series, vol. policy paper2. Available at: https://researchictafrica.net/wp/wp-content/uploads/2022/02/Digital-Taxation-can-it-contribute-to-more-just-resource-mobilisation-in-post-pandemic-reconstruction.pdf (Accessed 1 July 2023).

Onyango, S. 2022. 'Africa Accounts for 70% of the World's $1 trillion Mobile Money Market'. *Quartz*. Available at: https://qz.com/africa/2161960/gsma-70-percent-of-the-worlds-1-trillion-mobile-money-market-is-in-africa (Accessed 29 October 2023).

Otter, A. 2009. 'Telkom Owners in Line for Vodacom Bonus'. *Mybroadband*. https://mybroadband.co.za/news/telecoms/6928-telkom-owners-in-line-for-vodacom-bonus.html (Accessed 18 December 2015).

OUT-LAW.Com. 2015. 'South Africa's Vodacom Gets Green Light to Buy Neotel'. *OUT-Law*, 7 June. Available at: http://www.out-law.com/en/articles/2015/july/south-africas-vodacom-gets-green-light-to-buy-neotel/ (Accessed 1 August 2015).

Parke, P. 2016. 'More Africans Have Access to Cell Phone Service Than Piped Water'. *CNN*. https://edition.cnn.com/2016/01/19/africa/africa-afrobarometer-infrastructure-report/index.html#:~:text=In%20Africa%2C%20less%20than%20one,Africans%20have%20cell%20phone%20service (Accessed 18 December 2020).

Parliament of the Republic of South Africa. 2007. 'Report of the Committee on the ad hoc Review of Chapter 9 and Associated Institutions A'. Pretoria: Parliament of the Republic of South Africa.

Parliament of the Republic of South Africa. 2016. 'Amended Broad-Based Black Economic Empowerment (B-BBEE) Ict Sector Code'. Pretoria: Parliament of the Republic of South Africa.

Parliament of the Republic of South Africa. n.d. 'A People's Government, The People's Voice'. Available at: http://www.parliament.gov.za/live/content.php?Item_ID=296 (Accessed 14 February 2017).

Paterson, M., and Virk, K. 2014. 'South Africa's Foreign Policy after Two Decades'. Cape Town.

Paul, K. 1986. 'The Inadequacy of Sullivan Reporting'. *Business & Society Review*, (57), 61–5.

Pew Research Center. 2015. 'Cell Phones in Africa: Communication Lifeline'. Pew Research Centre: Global Attitudes & Trends, 15 April. Available at: http://www.pewglobal.org/2015/04/15/cell-phones-in-africa-communication-lifeline/ (Accessed 1 March 2016).

Pfister, R. 2005. *Apartheid South Africa and African States: From Pariah to Middle Power, 1962–1994*. New York: I.B. Tauris.

Pick, J. B., and Sarkar, A. 2015. *The Global Digital Divides: Explaining Change*. Berlin: Springer.

Pigman, G. A., and Deos, A. 2008. 'Consuls for Hire: Private Actors, Public Diplomacy'. *Place Branding and Public Diplomacy*, 4: 85–96.

Pitcher, A. 2012. 'Was Privatisation Necessary and Did It Work? The Case of South Africa'. *Review of African Political Economy*, 39(132), 243–60. Available at: http://www.jstor.org/stable/42003274 (Accessed 1 July 2013).

Pougala, J. -P. 2012. 'The Lies behind the West's War on Libya', in S. Ekine and F. Manji (eds), *African Awakening: The Emerging Revolutions*. Cape Town: Fahamu/Pambazuka, 170–9.

Premium Times. 2015. 'Bonga Oil Spill: Fishermen Seek Buhari's Help'. *Premium Times*, 19 April. Available at: http://www.premiumtimesng.com/news/

more-news/181706-bonga-oil-spill-fishermen-seek-buharis-help.html (Accessed 31 December 2015).

Prinsloo, L., and Bonorchis, R. 2016. 'PIC Slams R23.7m Pay Out for Former MTN Boss'. *Fin24*, 25 April. Available at: http://www.fin24.com/Tech/Companies/pic-slams-r237m-pay-out-for-former-mtn-boss-20160425 (Accessed 3 June 2016).

Procassini, A. A. 2000. 'Creating Cases in Commercial Diplomacy'. Commercial Diplomacy. Available at: http://www.commercialdiplomacy.org/manuals/manual_casestudies.htm (Accessed 28 June 2012).

Promotion of Bantu Self-Government Act No. 46 of 1959. Parliament of South Africa.

Provost, C., and Kennard, M. 2015. 'The Obscure Legal System that Lets Corporations Sue Countries'. *The Guardian*, 10 June. Available at: http://www.theguardian.com/business/2015/jun/10/obscure-legal-system-lets-corportations-sue-states-ttip-icsid (Accessed 28 October 2015).

Puhl, J. 2013. 'Africa's Transformative Digital Revolution'. *Spiegel Online*, 5 December. Available at: http://www.spiegel.de/international/world/silicon-savannah-how-mobile-phones-and-the-internet-changed-africa-a-936307.html (Accessed 1 February 2017).

Qiang, C. Z.-W. 2009. 'What Are the Economic and Social Impacts of the Mobile Phone Sector in Developing Countries'. *Proparcp Magazine*, (4).

Qobo, M. 2012. 'South Africa's Policy: An Assessment of the Draft White Paper and the ANC's Policy Document on International Relations'. Policy Brief No. 2, South African Foreign Policy Initiative.

Qobo, M. 2016. Interview Held at the University of Johannesburg (O. P. Ayodele, Interviewer).

Raffinetti, C. 2015. 'Licensing of Communications Networks and Services: Case Study of Market Liberalisation in South Africa and the United Kingdom'. *The African Journal of Information and Communication*, 2015(14), 50–61.

Ramamurti, R. 2000. 'A Multilevel Model of Privatization in Emerging Economies'. *The Academy of Management Review*, 25(3).

Rao, M. 2011. 'Mobile Africa Report 2011: Regional Hubs of Excellence and Innovation'. Mobile Monday Realities. London: Berghahn.

Reck, J., and Wood, B. 2003. *What Works: Vodacom's Community Services Phone Shops*. Washington, DC: World Resources Institute.

Rees, M. 2012. 'IDC to Invest in African Hotel Expansion'. *Moneyweb*. Available at: http://www.moneyweb.co.za/mw/view/mw/en/page295023?oid=568622&sn=2009+Detail&pid=295025 (Accessed 14 June 2012).

Republic of South Africa. 'Promotion of Bantu Self-Government Act 46 of 1959'.

Republic of South Africa. 'Republic of South Africa. Bantu Homelands Citizenship Act No. 26 of 1970'.

Russell, B. 2009. *Freedom and Organization,* 1st ed. London: Routledge.

Reuters. 2009. 'Nyanda: "MTN must remain African"'. *Fin24* Archives. Available at: http://www.fin24.com/Companies/Nyanda-MTN-must-remain-African-20090913 (Accessed 3 June 2016).

Reuters. 2012. 'Large Companies Avoiding Tax'. *Engineering News*. Available at: http://www.engineeringnews.co.za/print-version/large-companies-avoding-tax-2012-05-08 (Accessed 3 June 2012).

Reuters. 2016. 'South Africa to Quit International Criminal Court'. *The Guardian.* Available at: https://www.theguardian.com/world/2016/oct/21/south-africa-to-quit-international-criminal-court-document-shows (Accessed 3 March 2017).

Reuters. 2019. 'Under Siege in Nigeria, South African Businesses Shut Stores'. Available at: https://www.reuters.com/article/nigeria-safrica/under-siege-in-nigeria-south-african-businesses-shut-stores-idINKCN1VP154 (Accessed 4 August 2023).

Reuters and I-Net Bridge. 2010. 'MTN Reveals R8.1bn BEE Deal'. *FIN24* Archives. Available at: http://www.fin24.com/Companies/MTN-reveils-R81bn-BEE-deal-20100715 (Accessed 22 March 2016).

Rogers, M., and Pedros, X. 2017. 'Taxing Mobile Connectivity in Sub-Saharan Africa: A Review of Mobile Sector Taxation and Its Impact on Digital Inclusion'. GSMA. Available at: https://www.gsma.com/mobilefordevelopment/wp-content/uploads/2017/07/Taxing-mobile-connectivity-in-Sub-Saharan-Africa_July-2017.pdf (Accessed 3 July 2020).

Ronen, Y. 2011. *Transition from Illegal Regimes under International Law.* Cambridge: Cambridge University Press.

Rose, R. 2008. 'SA's Third Most Empowered Company, HCI'. Available at: https://www.hci.co.za/2008/04/04/here-empowerment-is-the-real-deal/ (Accessed 2 August 2023).

RSA. 1958. 'Post Office Act 44 of 1958'. Pretoria: Republic of South Africa.

RSA. 2016. 'Amended Broad-Based Black Economic Empowerment (B-BBEE) ICT Sector Code'. Available at: https://www.bbbeecommission.co.za/wp-content/uploads/2018/07/Amended-ICT-Sector-Code-40407.pdf (Accessed 4 August 2023).

Rubin, A. J., and de la Baume, M. 2014. 'Claims of French Complicity in Rwanda's Genocide Rekindle Mutual Resentment'. *New York Times*. Available at: http://

www.nytimes.com/2014/04/09/world/africa/claims-of-french-complicity-in-rwandas-genocide-rekindle-mutual-resentment.html?_r=0 (Accessed 31 March 2016).
Rumney, R. 2005. 'Who Owns South Africa: An Analysis of State and Private Ownership Patterns', in J. Daniel, R. Southall and J. Lutchman, *State of the Nation: South Africa 2004–2005*. Cape Town: HSRC Press.
Russett, B. M. 1974. *Power and Community in World Politics*. San Francisco: W.H. Freeman.
Sachs, E. S. 1957. *Garment Workers in Action: History of the Garment Workers of South Africa to 1952*. Johannesburg: Eagle Press.
SADC Overview. n.d. 'South African Development Community'. Available at: http://www.sadc.int/about-sadc/overview/ (Accessed 3 March 2013).
Salmond, J. 1997. 'Swaziland: I – of Trade Unions and Transformation'. *Southern Africa Report*, 12(3), 5–8.
Samuel, C. 2013. 'The Dark Side of Foreign Direct Investment: A South African Perspective'. SAIIA Occasional Paper No. 167 (December). Johannesburg: South African Institute of International Affairs.
Sánchez, D. 2008. 'Transnational Telecommunications Capital Expanding from South Africa into Africa: adapting to African Growth and South African Transformation Demands'. *African Sociological Review/Revue Africaine de Sociologie*, 12(1), 105–23.
Scholvin, S., and Draper, P. 2012. 'The Gateway to Africa? Geography and South Africa's Role as an Economic Hinge Joint between Africa and the World'. *South African Journal of International Affairs*, 19(3), 381–400.
Schrire, R. 1982. 'The Homelands: Political Perspectives', in R. Schrire (ed.), *South Africa: Public Policy Perspectives*. Cape Town: Juta.
'Second Anglo-Boer War – 1899–1902'. n.d. *South African History Online*. Available at: https://www.sahistory.org.za/article/second-anglo-boer-war-1899-1902 (Accessed 31 July 2023).
Seekings, J. 2008. 'The Carnegie Commission and the Backlash against Welfare State-Building in South Africa, 1931–1937'. *Journal of Southern African Studies*, 34(3), 515–37.
Sell, K. S. 2003. *Private Power, Public Law: The Globalization of Intellectual Property Rights*. Cambridge: Cambridge University Press.
Seme, P. 1906. 'Speech by Pixley Seme the Regeneration of Africa 5 April 1906'. Available at: https://www.sahistory.org.za/sites/default/files/regeneration_of_africa_by_Pixley_Seme.pdf (Accessed 1 July 2023).
Sergeant, M. 1993a. 'Cellphone Telephone Industry Shapes Up'. *Business Day*, 9 June, p. 1.

Sergeant, M. 1993b. 'Phone Controversy Rages'. *Business Day*, 30 July, p. 3.

Serobe, G. 2015. 'It's Time for South African Business to Step Up'. Mckinsey, June. Available at: http://www.mckinsey.com/global-themes/middle-east-and-africa/its-time-for-south-african-business-to-step-up (Accessed 2 August 2016).

Sethi, S. P. 1991. 'The Prospects and Problems of Two-Way Exchange in the Economic Arena between Post-Apartheid South Africa and the Rest of Africa', in *The Challenges of Post-Apartheid South Africa: Conclusions and Papers Presented at a conference of the African Leadership Forum*. Windhoek: African Leadership Forum.

Seyle, C. 2016. 'How Business Can Support the Responsibility to Protect'. OEF Policy Brief. https://oneearthfuture.org/sites/default/files/documents/publications/The_Role_of_Business_R2P-Digital.pdf.

Sharma, A., and Bellman, E. 2009. 'Bharti, MTN Disconnect Deal Worth $24 Billion'. *Wall Street Journal*. Available at: http://www.wsj.com/articles/SB125431578653952371 (Accessed 4 June 2016).

Sharman, J. C., and Phillips, A. 2020. *Outsourcing Empire: How Company-States Made the Modern World*. Princeton, NJ: Princeton University Press.

Shezi, L. 2017. 'The Siege on Internet Freedom: How African Governments Are Tightening Their Grips'. *htxt.africa*. Available at: http://www.htxt.co.za/2017/01/31/the-siege-against-internet-freedom-iron-fist-african-governments-tighten-their-grips/ (Accessed 1 February 2017).

Siegal, E. 2009. 'What Does Torque in a Car Do?' *Science Blogs*. Available at: http://scienceblogs.com/startswithabang/2009/04/21/what-does-torque-in-a-car-do/ (Accessed 15 October 2015).

Sieh, R. D. 2016. 'Liberia: Bankruptcy Fears? Lonestarcell/MTN Cautions Shareholders'. *All Africa*. Available at: http://allafrica.com/stories/201601251307.html (Accessed 31 March 2016).

Sigonyela, M. 2001. 'South Africa: Comrades Slug It Out for MTN Shares'. *AllAfrica*. Available at: http://allafrica.com/stories/200108080093.html (Accessed 20 March 2016).

Siko, J. 2014. *Inside South Africa's Foreign Policy: Diplomacy in Africa from Smuts to Mbeki*. New York: I.B. Tauris.

Skålnes, T. 1997. 'Institutions, Interest Groups and Economic Politics', in S. Lindberg and A. Svarrissson (eds), *Social Movements in Development: The Challenge of Globalization and Democratization*. Basingstoke: Macmillan Press.

Sklair, L. 2000. 'The Transnational Capitalist Class and the Discourse of Globalization'. *Cambridge Review of International Affairs*, 14(1), 67–85.

Sklair, L. 2002. 'Democracy and the Transnational Capitalist Class'. *Annals of the American Academy of Political and Social Science*, 581, 144–57.

Söderbaum, F. 2004. 'Modes of Regional Governance in Africa: Neoliberalism, Sovereignty Boosting, and Shadow Networks'. *Global Governance*, 10(4), 419–36.

Soko, M., and Belchin, N. 2009. 'South Africa's Policy towards Zimbabwe: A Nexus between Foreign Policy and Commercial Interests'. *South African Journal of International Affairs*, 16(1), 33–48.

South African History Online. 2011. 'South Africa Concludes an Agreement with the Marconi Company for a High'. *South African History Online*. Available at: https://www.sahistory.org.za/dated-event/south-africa-concludes-agreement-marconi-company-high (Accessed 14 August 2018).

South African Institute for Race Relations. 2011. '2009/2010 South Africa Survey'. Johannesburg: South African Institute for Race Relations.

Southall, R. 1994. 'The New South Africa in the New World Order: Beyond the Double Whammy'. *Third World Quarterly*, 15(1), 33–48.

Southall, R. 2006. 'The Logic of Black Economic Empowerment (No. 2006: 28)'. DIIS Working Paper. https://www.researchgate.net/publication/242508684_THE_LOGIC_OF_BLACK_ECONOMIC_EMPOWERMENT (Accessed 18 December 2015).

Speckman, A. 2015. 'Black Investors Turn Activist at MTN'. *Sunday Times Business*. Available at: http://www.timeslive.co.za/sundaytimes/businesstimes/2015/05/31/Black-investors-turn-activist-at-MTN (Accessed 20 June 2015).

Spicer, M. 2016. 'The Business Government Relationship: What Has Gone Wrong?' *Focus*, (78). https://hsf.org.za/publications/focus/focus-78-the-economy-1/michael-spicer.pdf (Accessed 18 December 2020).

Spillane, C., Alexander, G. S., and Antony, A. 2013. 'MTN Deal Talks with Reliance Communications Said Stalled'. London: Bloomberg. Available at: http://www.bloomberg.com/news/articles/2013-07-09/mtn-deal-talks-with-reliance-communications-said-stalled (Accessed 4 June 2016).

Staff Reporter. 2007. 'Mbeki Critical of Crime Issues in APRM Report'. *Mail & Guardian*. Available at: http://mg.co.za/article/2007-12-06-mbeki-critical-of-crime-issues-in-aprm-report (Accessed 20 January 2017).

Staff Reporter. 2009. 'Icasa: MTN, Bharti Deal May Need Public Hearings'. *Mail & Guardian*. Available at: http://mg.co.za/article/2009-09-14-icasa-mtn-bharti-deal-may-need-public-hearings (Accessed 3 June 2016).

Staff Writer. 2008. 'South Africa's Economy Shrinks by 1.2%, Raising Prospect of Recession'. *Sunday Times Business*. Available at: http://www.timeslive.co.za/sundaytimes/businesstimes/2016/06/08/South-Africas-economy-shrinks-by-1.2-raising-prospect-of-recession (Accessed 20 January 2017).

Staff Writer. 2022. 'Vodacom in Multibillion-Rand B-BBEE Drive'. *ITWeb*. Available at: https://www.itweb.co.za/content/GxwQDM1DmBx7lPVo (Accessed 5 August 2023).

Statista. 2023. 'Africa: Mobile Phone Ownership among Youth'. *Statista*. Available at: https://www.statista.com/statistics/1266162/young-people-s-ownership-rate-of-mobile-phones-in-africa/ (Accessed 29 October 2023).

Stern, M., and Ramkolowan, Y. 2021. 'Understanding South Africa's Trade Policy and performance, Working Papers WP/20/17'. Available at: https://ideas.repec.org/p/rbz/wpaper/11017.html (Accessed 30 October 2023).

Stern, P. J. 2015. 'The English East India Company and the Modern Corporation: Legacies, Lessons, and Limitations'. *Seattle UL Rev*, 39(423), 423–45.

Stevenson, D. 2008. 'Investors Chronicle – Magazine and Web Content: Buy into Africa'. *Investors Chronicle*, 15 August. Available at: https://www.investorschronicle.co.uk (Accessed 4 August 2023).

Stewart, J. B. 2011. 'Amandla! The Sullivan Principles and the Battle to End Apartheid in South Africa 1975–1987'. *Journal of African American History* 96(1), 62–89.

Steytler, M., and Mackay, G. 2015. 'Social Protests on the Rise in South Africa'. Press Release: Institute of Race Relations. Available at: http://irr.org.za/reports-and-publications/media-releases/social-protests-on-the-rise-in-south-africa (Accessed 20 October 2015).

Stuurman, K. 2004. 'Intervention in Africa? The Mbeki Presidency's Role in Changing the OAU', in *African Renewal, African Renaissance: New Perspectives on Africa's Past and Africa's Present*. The African Studies Association of Australia and the Pacific (AFSAAP) Annual Conference, University of Western Australia, 26–28 November.

Sullivan, N. P. 2007. *You Can Hear Me Now: How Micro Loans and Cell Phones Are Connecting the World's Poor to the Global Economy*. San Francisco: Jossey-Bass.

Susman Associated Press. 1993. 'ANC Hits S. African Phone Deal – Flap Could Deter Return of Foreign Investment'. *Daily News of Los Angeles (CA)*, 26 September, p. U1.

Sutherland, E. 2010–11. 'International Mobile Roaming: Progress and Challenges in African Markers'. *The African Journal of Information and Communication* (11), 2–22.

Sutherland, E. 2012. 'Corruption in Telecommunications: Problems and Remedies'. *Info*, 14(1), 4–19.

Sutherland, E. 2014a. 'Mobile Telecommunications in Africa: Issues for Business, Government & Society'. *Government & Society (January 26, 2014)*. https://dx.doi.org/10.2139/ssrn.2374346.

Sutherland, E. 2014b. 'Telecommunications in the Kingdom of Swaziland'. *Communicatio*, 40(4), 376–99.

Sutherland, E. 2015. 'MTN: A South African Mobile Telecommunications Group in Africa and Asia'. *Communicatio*, 41(4).

Swazi Media Commentary. 2015. 'Swaziland: MTN 'Keeps Swazi King in Its Pocket'. *AllAfrica*. Available at: http://allafrica.com/stories/201502041457.html (Accessed 22 September 2015).

Taal, M. 2014. 'South African Multinational Corporations in Africa – Trends in 2013/2014'. Johannesburg: Labour Research Services. Friedrich Ebert Stiftung.

Tafirenyika, M. 2014. 'Intra-Africa Trade Hurdles: Beyond Politics'. *IOL Business*. Available at: http://www.mobi.iol.co.za/#!/article/intra-africa-trade-hurdles-beyond-politics-1.174149 (Accessed 1 October 2015).

Taliaferro, J. W. 2012. 'Neoclassical Realism and the Study of Regional Order', in T. V. Paul (ed.), *International Relations Theory and Regional Transformation*. Cambridge: Cambridge University Press.

Tangri, R. 1999. *The Politics of Patronage in Africa: Parastatals, Privatization & Private Enterprise*. London: Africa World Press.

Tarrant, H. 2009. 'MTN/Bharti Deal Off: Parties Decide to Walk Away'. *Moneyweb*. Available at: http://www.moneyweb.co.za/archive/mtnbharti-deal-off/ (Accessed 3 June 2016).

Tarrant, H. 2015. M-Pesa's a Bigger Flop in SA Than Vodacom's Letting on'. *Moneyweb*. Available at: http://www.moneyweb.co.za/moneyweb-opinion/m-pesas-a-bigger-flop-in-sa-than-vodacoms-letting-on/ (Accessed 20 August 2015).

Tawfik, R. 2010. 'NEPAD and African Development: Towards a New Partnership between Development Actors in Africa'. *African Journal of International Affairs*, 11(1), 55–70.

te Velde, W., and Leftwich, A. 2010. 'State-Business Relations and Economic Growth in Sub-Saharan Africa'. Consortium on Improving Institutions for Pro-Poor Growth (IPPG). Available at: https://assets.publishing.service.gov.uk/media/57a08b0f40f0b652dd000a94/8851_State-Business-Relations-Final.pdf (Accessed 1 November 2011).

Technology Innovation Agency Act 26 of 2008. Republic of South Africa.

Technology Times Founder. n.d. 'Mobile Networks Are the New "One-Stop-Shop" Businesses'. *Technology Times*. Available at: https://www.tia.org.za/core/uploads/2021/06/Technology_Innovation_Agency_Act_No._26_of_2008_1-April-2021-003.pdf (Accessed 1 August 2015).

TeleGeography. 2016. 'Cellsaf Challenges Cell C's Restructuring Plan'. Available at: https://www.commsupdate.com/articles/2016/11/30/cellsaf-challenges-cell-cs-restructuring-plan/ (Accessed 28 July 2023).

Terreblanche, S. 2005. 'A Review of the Rise of the Afrikaner Middle Class in South Africa: A Review of the Rise of the Afrikaner Middle Class in South Africa'. Available at: https://www.ekon.sun.ac.za/sampieterreblanche/wp-content/uploads/2020/07/2005-A-review-of-the-rise-of-the-Afrikaner-middle-class-in-South-Africa.pdf (Accessed 26 July 2023).

The Ministry for Posts, Telecommunications and Broadcasting. 1996. White Paper on Telecommunications Policy. https://www.gov.za/sites/default/files/gcis_document/201610/40325gon1212.pdf (Accessed 18 February 2019).

The Republic of South Africa. 2003. No. 53 of 2003: Broad-Based Black Eco nomic Empowerment Act. https://www.gov.za/sites/default/files/gcis_document/201409/a53-030.pdf (Accessed 18 December 2015).

The Tony Elumelu Foundation. n.d. 'Driving Africa's Renaissance thr ough Africapitalism'. The Tony Elumelu Foundation. Available at: http://tonyelumelufoundation.org/africapitalisminstitute/driving-africas-renaissance-africapitalism/ (Accessed 28 October 2015).

ThisDay LIVE. 2017 'African Govts Challenge ITU over Hosting Rights'. *THISDAYLIVE*. Available at: https://www.thisdaylive.com/index.php/2017/09/28/african-govts-challenge-itu-over-hosting-rights/ (Accessed 18 December 2020).

Thompson, R. J., and Nicholls, B. M. 1993. 'The Glen Grey Act: Forgotten Dimensions in an Old Theme'. *South African Journal of Economic History*, 8(2), 58–70.

Thomson, A., and Spillane, C. 2014. 'Vodafone Gets Chance to Tighten Grip on Africa'. *Bloomberg Business*. Available at: http://www.bloomberg.com/news/articles/2014-10-15/vodafone-could-tighten-grip-on-africa-as-south-africa-seeks-exit (Accessed 8 October 2015).

Tijani, M. 2016. 'MTN: We contribute 4.5% of Nigeria's GDP'. *The Cable*. Available at: https://www.thecable.ng/regulatory-pressure-severely-impacting-us-say-mtn-nigeria (Accessed 3 January 2017).

Tjemolane, T., Neethling, T., and Schoeman, A. 2012. 'South Africa's Foreign Policy and Africa: Continental Partner or Hegemon?' *Africa Review*, 4(2): 87–106.

Tomaselli, K. G. 2000. South African Media 1994–7: Globalising via Political Economy', in J. Curran and M.-J. Park (eds), *De-Westernizing Media Studies*. London: Routledge, 247–59.

Trahar, A. J. 2002. 'Africa –a Continent of Opportunity'. *Anglo American PLC*. Available at: http://www.angloamerican.com/~/media/Files/A/Anglo-Ameri

can-Plc/media/speeches/2002sp/2002-01-28/2002-01-28.pdf (Accessed 28 October 2015).

Transparency International. 2015. 'What Is Corruption?'. *Transparency International*. Available at: http://www.transparency.org/what-is-corruption/#define (Accessed 31 March 2016).

Tshimologong. 2016. 'About'. Available at: http://tshimologong.joburg/ (Accessed 1 February 2017).

Tyamzashe, M. 2007. 'The Role of Business in Promoting South Africa's Foreign Policy: "A Vodacom Group (PTY) LTD perspective"', in *The Role of Business in Promoting South Africa's Foreign Policy Ideals*. Cape Town: Seminar to Parliament Portfolio Committee on International Relations. https://pmg.org.za/committee-meeting/8511/.

Ubabukoh, O. 2015. 'Ericsson's Engineer Installed Biafra's Transmitters on MTN's Masts'. *Punch*. Available at: http://www.punchng.com/ericssons-engineer-installed-biafras-transmitters-on-mtns-masts/ (Accessed 31 March 2016).

Ujuh. 2015. 'BEE Scheme, MTN Zakhele, Receives a Fat Dividend'. *TechFinancials*. Available at: http://techfinancials.co.za/2015/05/30/bee-scheme-mtn-zakhele-receives-a-fat-dividend/ (Accessed 9 September 2015).

UN. 1947. Economic and Social Council (5th session: 1947: Lake Success, NY), E/RES/90(V). https://digitallibrary.un.org/record/561096?ln=en&v=pdf (Accessed 18 December 2015).

UNCTAD. 2012. 'Thirteenth Session African Trade Ministers Meeting UNCTAD XIII Pre-Conference Event'. United Nations Conference on Trade and Development Secretariat. https://unctad.org/system/files/official-document/td464_en.pdf (Accessed 18 December 2015).

UNCTAD. 2015. 'World Investment Report: Reforming International Investment Governance'. Geneva: United Nations Conference on Trade and Development.

UNDP. 2012. 'Mobile Technologies: Enhancing Human Development through Participation and Innovation'. Primer. United Nations Development Fund.

UNDP General Council. 1982. 'Other Funds and Programmes: Technical Cooperation among developing countries'. Interim Report of the Administrator [Geneva], DP/1982/42.Add.I, 16 April. Available at: http://web.undp.org/execbrd/archives/sessions/gc/29th-1982/DP-1982-42-Add1.pdf (Accessed 1 April 2012).

UNESCO. 2011. 'Communication and Information Activities'. The United Nations Educational, Scientific and Cultural Organization. Available at: http://portal.unesco.org/ci/en/ev.php-URL_ID=3038&URL_DO=DO_TOPIC&URL_SECTION=201.html (Accessed 22 April 2015).

UNGA. 1974. 'Declaration for the Establishment of a New International Economic Order: A/RES/S-6/3201'. 1 May. https://digitallibrary.un.org/record/218450/files/A_RES_3201%28S-VI%29-EN.pdf?ln=en (Accessed 18 December 2015).

United Nations Economic Commission for Africa (UNECA). 2017. 'Towards Improved Ac cess to Broadband in Africa'. Addis Ababa: UNECA.

Vale, P. C. 1982. 'South Africa and the Changing International Community: Interpreting the Future in the Light of the Past', in R. Schrire (ed.), *South Africa: Public Policy Perspectives*. Cape Town: Juta, 629–34.

Vale, P., and Taylor, P. 1999. 'South Africa's Post-Apartheid Foreign Policy Five Years on: From Pariah State to Just Another Country'. *The Round Table*, 88(352).

Van Bergeijk, P. A., and Moons, S. J. 2011. 'A Meta Analysis on the Effect of Economic Diplomacy on International Economic Flows'. Copenhagen: ETSG Conference.

van der Westhuizen, C. 2012. 'South Africa: Economic Growth, Poverty and Inequality'. Foresight Africa: Top Priorities for the Continent. https://www.brookings.edu/wp-content/uploads/2016/06/01_snapshot_south_africa_westhuizen.pdf (Accessed 18 December 2015).

Van Grasdorff, E. 2005. *African Renaissance and Discourse Ownership in the Information Age: The Internet as a Factor of Domination*. Munster: LIT.

van Zyl, G. 2015a. 'MTN Hit with R9.3m Fine in Uganda'. *Sunday Times*. Available at: http://www.timeslive.co.za/sundaytimes/businesstimes/2015/11/19/MTN-hit-with-R9.3m-fine-in-Uganda (Accessed 31 March 2016).

van Zyl, G. 2015b. SA Fails on Digital Migration'. *Mail & Guardian* Online. Available at: http://mg.co.za/article/2015-06-17-sa-fails-on-digital-migration (Accessed 19 August 2015).

Vanham, P. 2015. 'Why Is Angola's Capital the Most Expensive City in the World?'. World Economic Forum. Available at: https://www.weforum.org/agenda/2015/06/why-is-angolas-capital-the-most-expensive-city-in-the-world/ (Accessed 1 March 2016).

Veneranda, S. 2015. 'Vodacom Tanzania CEO Quits'. *The East African*. Available at: http://www.theeastafrican.co.ke/business/vodacom-tanzania-ceo-quits/-/2560/business/vodacom (Accessed 1 September 2015).

Venter, D. 2001. South African Foreign Policy Decision Making in the African Context', in G. M. Khadiagala and T. Lyons (eds), *African Foreign Policies: Power and Process*. Boulder: Lynne Rienner.

Ventures Africa. 2015. 'Vodacom Sinks 100m in Tanzania for Rural Expansion'. *Ventures Africa*. Available at: http://venturesafrica.com/vodacom-sinks-100m-in-tanzania-for-rural-expansion/ (Accessed 18 September 2015).

Vermeulen, D. J. 1998. 'The Remarkable Dr. Hendrik van der Bijl'. *Proceedings of the IEEE*, 86(12), 2445–54.

Versi, A. 2012. 'Transforming a Continent'. African Business. https://african.business/2012/05/economy/transforming-a-continent (Accessed 18 December 2015).

Vickers, B., and Ajulu, C. 2008. 'South Africa's Economic Diplomacy: Trade and Investment Promotion'. Report Prepared for The Presidency's Fifteen Year Review, Institute for Global Dialogue.

Vickers, B., and Cawood, R. 2017 'South Africa's Corporate Expansion: Towards an "SA Inc." Approach in Africa', in A. Adebajo and K. Virk (eds), *Foreign Policy in Post-Apartheid South Africa*. New York: I.B. Tauris.

Vodacom. 2022. 'Vodacom Group Limited Integrated Report for the Year Ended 31 March 2022'. Available at: https://vodacom-reports.co.za/integrated-reports/ir-2022/documents/Our-strategy.pdf (Accessed 4 August 2023).

Vodacom. 2023. 'Vodacom Group Limited Fact Sheet 2023'. Available at: https://vodacom.com/pdf/about-us/fact-sheet/2023/vodacom-group-limited-fact-sheet-2023-new.pdf (Accessed 25 November 2023).

Vodacom Group. 2022. 'Vodacom Achieves Level 1 B-BBEE Status, News Releases'. https://www.vodacom.com/news-article.php?articleID=7786 (Accessed 18 March 2024).

Vodafone. 2018. 'Change in Vodafone's Ownership of Vodacom Group and Vodacom Group's Ownership of Its South African Subsidiary'. Vodafone.com. Available at: https://www.vodafone.com/news/inclusion/vodacom-ownership (Accessed 4 August 2023).

Vodafone Group Plc. 1995. 'Annual Report and Accounts for Year Ended March 1995'. Available at: https://www.vodafone.com/content/dam/vodcom/files/investors/annual-report/earlier/annual_report_accounts_1995.pdf (Accessed 3 August 2023).

Vodafone Group Plc. 2003. 'Annual Report & Accounts and Form 20-F For the Year Ended 31 March 2003'. Available at: https://www.vodafone.com/content/dam/vodcom/files/investors/annual-report/earlier/annual_report_accounts_2003.pdf (Accessed 3 August 2023).

'Vodafone in South Africa'. 2008. *Financial Times*, 3 June. Available at: https://www.ft.com/content/1ee7d198-30ad-11dd-bc93-000077b07658 (Accessed 2 August 2023).

Wambua, P. W., and Balongo, S. 2016. 'Building on Progress: Infrastructure Development Still a Major Challenge in Africa'. Afrobarometer Dispatch No 60. Available at: http://afrobarometer.org/sites/default/files/publications/Dispatches/ab_r6_dispatchno69_infrastructure_remains_challenge_en.pdf (Accessed 14 February 2017).

Warschauer, M. 2003. 'Dissecting the "Digital Divide": A Case Study in Egypt'. *The Information Society: An International Journal*, 19(4), 297–304.

Warzui, B. 2010. 'Using Mobile Phones for Monitoring Human Rights Violations in the DRC', in S. Ekine (ed.), *SMS Uprising: Mobile Phone Activism in Africa*. Cape Town: Pambazuka Press, 138–42.

Waverman, L., Meschi, M., and Fuss, M. 2005. 'The Impact of Telecoms on Economic Growth in Developing Countries'. The Vodafone Policy Paper Series (3): 10–24. https://www.vodafone.com/content/dam/vodcom/sustainability/pdfs/Vodafone%20Survey.pdf.

Weingrod, A. 1968. 'Patrons, Patronage, and Political Parties'. *Comparative Studies in Society and History*, 10(4), 377–400.

Weis, L. 1998. *The Myth of the Powerless State: Governing the Economy in a Glob al Area*. Cambridge: Polity Press.

Whalley, J., and Curwen, P. 2013. 'Unravelling Complex Organisational Structures among Mobile Operators'. *Info*, 15(4), 3–22.

White, L. 2016. Interview held at Gordon Institute of Business Science, Illovo. (O. Ayodele, Interviewer).

Wickman, M. (no date). 'South Africa – Ericsson'. Available at: https://www.ericsson.com/en/about-us/history/places/africa/south-africa (Accessed 7 August 2023).

Wild, S. 2013. 'Confusion Gives Technology Innovation Agency Trouble'. *Mail & Guardian*. Available at: http://mg.co.za/article/2013-07-11-troubled-technology-innovation-agency-left-spinning (Accessed 3 June 2016).

Wild, S. 2016. 'Innovation Agency CEO Works on Bringing Stability'. *Business Day*, 29 August.

Williams, P. 2014. 'Africans Are Doing It for Themselves with Great Success'. *SAnews*. Available at: http://www.sanews.gov.za/south-africa/africans-are-doing-it-themselves-great-success (Accessed 25 March 2016).

Winds of Change Speech'. no date. 'Hendrik Verwoerd's Response to the 'Winds of Change' Speech | South African History Online'. https://www.sahistory.org.za/archive/hendrik-verwoerds-response-winds-change-speech (Accessed 18 December 2015).

World Bank. 2013. 'State of the Poor'. World Bank. Available at: http://www.worldbank.org/content/dam/Worldbank/document/State_of_the_poor_paper_April17.pdf (Accessed 10 October 2015).

World Bank. 2013. 'Mobile Usage at the Base of the Pyramid: Research Findings from Kenya and South Africa'. The World Bank. https://openknowledge.worldbank.org/bitstreams/51f9248e-ce26-57e6-a594-51886792561f/download (Accessed 18 December 2015).

World Bank. 2015. 'Data'. World Bank. Available at: http://data.worldbank.org/country/congo-dem-rep?display=graph (Accessed 18 October 2015).

World Bank. 2023. 'Africa Development Indicators'. World Bank Data. https://databank.worldbank.org/source/africa-development-indicators (Accessed 1 December 2023).

World Bank Group. 2016a. 'Global Economic Prospects'. Regional Integrations and Spillovers. Available at: https://www.worldbank.org/content/dam/Worldbank/GEP/GEP2016a/Global-Economic-Prospects-January-2016-Spillovers-SSA.pdf (Accessed 28 August 2016).

World Bank Group. 2016b. 'Overview'. World Bank. Available at: http://www.worldbank.org/en/country/southsudan/overview (Accessed 29 July 2016).

World Bank/OECD. 2023. 'GDP (current US$) – Equatorial Guinea, Malawi, Namibia, Sao Tome and Principe, World Bank Open Data'. Available at: https://data.worldbank.org/indicator/NY.GDP.MKTP.CD?locations=GQ-MW-NA-ST (Accessed 30 October 2023).

Wray, R. 2007. 'Vodafone in Talks to Buy Control of South African Joint Venture'. *The Guardian*. Available at: https://www.theguardian.com/business/2007/sep/04/8 (Accessed 2 August 2023).

Wren, C. S. 1990. 'Turmoil Spreads to 2D "Homeland"'. Special to *New York Times*, 8 March.

Wroblewska, A. B. 2014. 'African Expansion Critical For South African Companies'. AFKI Original. Available at: http://afkinsider.com/73863/african-expansion-critical-south-african-companies/#sthash.fpq9wdhd.dpuf (Accessed 20 January 2017).

WTO. 1996. 'Negotiating Group on Basic Telecommunications'. World Trade Organization. Available at: https://www.wto.org/english/tratop_e/serv_e/telecom_e/tel23_e.htm (Accessed 4 June 2016).

Yebo Yethu. n.d. 'YeboYethu Limited'. Available at: https://www.yeboyethu.co.za/profile.php (Accessed 5 August 2023).

Zibi, S. 2015. 'UNEMBARGOED: Greek Tragedy's Moral Lesson for SA'. *Business Day BDlive*. Available at: http://www.bdlive.co.za/opinion/columnists/2015/07/06/unembargoed-greek-tragedys-moral-lesson-for-sa (Accessed 20 August 2015).

Zondi, S. 2013. 'Global Mandela: A Complexity'. *South African Foreign Policy Initiative Policy Brief No56*.

Zounmenou, D. D. 2014. Interview Held at the ISS Pretoria Office. (O. P. Ayodele, Interviewer).

Zuma, J. G. 2016. 'Intervention by His Excellency, President Jacob Zuma at the BRICS Leaders' Plenary Session, Goa, India'. South African Government. Available at: http://www.gov.za/speeches/president-jacob-zuma-brics-leaders%E2%80%99-plenary-session-16-oct-2016-0000 (Accessed 6 January 2017).

Index

Note: Figures are indicated by page number followed by 'f'. Tables are indicated by page number followed by 't'. Footnotes are indicated by the page number followed by 'n' and the footnote number e.g., 20 n.1 refers to footnote 1 on page 20.

www.ingramcontent.com/pod-product-compliance
Ingram Content Group UK Ltd.
Pitfield, Milton Keynes, MK11 3LW, UK
UKHW020815290926
15470UKWH00003B/1179

9 781350 411777